Rewilding the Urban Frontier

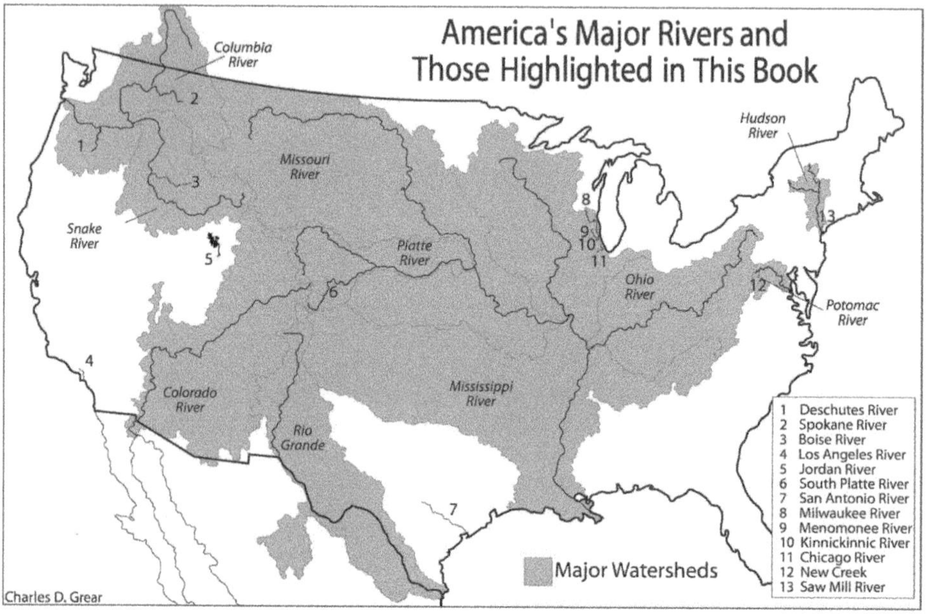
America's Major Rivers and
Those Highlighted in This Book
Columbia River
Missouri River
Snake River
Platte River
Ohio River
Hudson River
Potomac River
Colorado River
Rio Grande
Mississippi River
Major Watersheds
1 Deschutes River
2 Spokane River
3 Boise River
4 Los Angeles River
5 Jordan River
6 South Platte River
7 San Antonio River
8 Milwaukee River
9 Menomonee River
10 Kinnickinnic River
11 Chicago River
12 New Creek
13 Saw Mill River
Charles D. Grear

Rewilding the Urban Frontier

River Conservation in the Anthropocene

Edited by Greg Gordon

University of Nebraska Press | Lincoln

An earlier version of chapter 1, "Haunted," by Heidi Lasher, appeared as "Seeing Ghosts," in "The Nature Issue," *Litro Magazine*, October 28, 2021, https://www.litromagazine.com/usa/2021/10/seeing-ghosts/.

The University of Nebraska Press is part of a land-grant institution with campuses and programs on the past, present, and future homelands of the Pawnee, Ponca, Otoe-Missouria, Omaha, Dakota, Lakota, Kaw, Cheyenne, and Arapaho Peoples, as well as those of the relocated Ho-Chunk, Sac and Fox, and Iowa Peoples.

∞

Library of Congress Cataloging-in-Publication Data

Names: Gordon, Greg, 1963- editor.
Title: Rewilding the urban frontier : river conservation in the anthropocene / edited by Greg Gordon.
Description: Lincoln: University of Nebraska Press, [2024] | Includes bibliographical references and index.
Identifiers: LCCN 2023056384
ISBN 9781496230614 (hardback)
ISBN 9781496239921 (epub)
ISBN 9781496239938 (pdf)
Subjects: LCSH: Stream conservation—United States. | Stream ecology—United States. | Urban ecology (Biology)—United States. | Urban ecology (Sociology)—United States. | BISAC: NATURE / Ecosystems & Habitats / Rivers | HISTORY / United States / State & Local / West (AK, CA, CO, HI, ID, MT, NV, UT, WY)
Classification: LCC QH76 .R49 2024 | DDC 333.91/62160973—dc23/eng/20240416
LC record available at https://lccn.loc.gov/2023056384

Designed and set in Minion Pro by L. Welch.

Frontispiece: Major rivers of the United States and the rivers highlighted in this book. Created by Charles David Grear, used by permission.

Contents

Illustrations

Rewilding the Urban Frontier

Introduction

Origins

Greg Gordon

This is a story about a river, or rather, many rivers. It could be any river, really. It could even be your river. This river was once chock full of salmon—fish so abundant that Native peoples of many tribes once gathered by the hundreds to harvest fish, hold ceremonies, conduct marriages, exchange goods, and forge friendships and alliances.

Not only did the river contain fish, but it also housed beavers, especially in the tributaries. Beavers, in turn, attracted fur trappers, who, while transient, radically changed the ecosystem by removing said beavers. Other Euro-Americans recognized the power and opportunity of falling water and built sawmills while staking out townsites. In the broad valleys, settlers tapped the rivers for irrigation, building canals and diversions to water crops and orchards.

Railroads followed the river valleys and conveniently placed their switching yards and warehouses in the middle of towns near the river. Dependent upon rail transportation, other industries followed. With the advent of electrical power at the dawn of the twentieth century, investors built dams across the narrow gaps in the canyon for hydroelectric plants. These dams dramatically altered the river's hydrology, ended the salmon runs, and trapped toxic waste from upstream mines.

By the mid-twentieth century, the river had become an open sewer. The city's industrial core featured mills, factories, warehouses, stockyards, and rendering plants, all strung out along the river, flushing industrial waste and toxins from mining, agriculture, industry, and municipal systems.

The 1970s marked a dramatic environmental turn as people began to recognize the value of the river and initiated ambitious cleanup programs. The city remade the industrial downtown, ripped up rail yards, tore down warehouses, and created public parks and walkways. Before long the river was a source of civic pride. With the help of the 1972 Clean Water Act and other environmental laws, the public began to push for better water quality, habitat protection, and recreational access.

If any of this sounds familiar, that's because it's the same story across much of the United States. Urban rivers reflect our changing relationship with the natural world, from abundance, to abuse and exploitation, to neglect, to beautification. While this river story has generally been a narrative focused on human values, a biocentric view is now emerging. Ushering in a new era of river restoration, in 1992 Congress passed legislation to remove two dams on the Elwha River in Olympic National Park, Washington. Twenty years later, these dams finally came down, and the Elwha has largely returned to its presettlement conditions with braided channels, active sediment transport, and, most importantly, the return of salmon runs.

While on the East Coast in 1997, the Federal Energy Regulatory Commission—the agency in charge of relicensing dams—ordered the removal of the Edwards Dam on the Kennebec River, the first time the agency had taken such action. Two years later, the dam came down, and the river's aquatic life quickly rebounded with Atlantic sturgeon, alewives, and even seals, swimming up from the ocean. Water quality improved as well, and the residents of Augusta, Maine, rediscovered their river. The age of urban river restoration had begun.[1]

In addition to the physical restoration of hydrological function through dam removal, the twenty-first century has witnessed significant citizen engagement with local watersheds and a recognition of the importance of aquatic biodiversity, urban revitalization, and the implementation of forward-thinking legislation (such as the Clean Water Act). All of these factors have coalesced to begin the recovery of America's rivers.

At the same time, the challenges for recovery are daunting. America's rivers have been dammed, diverted, channelized, dewatered, and polluted. In short, urban rivers are the most abused, neglected, and damaged ecosystems in the country. And yet, the wildest places in the city tend to be along these same rivers and creeks that offer crucial wildlife corridors and connectivity to core conservation areas, such as state parks and national forests. Ironically, our postindustrial urban landscapes offer significant potential for ecological recovery and could potentially forestall the impending ecological crisis.

At the breaching of the Edwards Dam in 1999, Bruce Babbitt, then secretary of the Interior and a river advocate who proudly presided over the dismantling of multiple dams, stated, "Rivers flow through our lives, they flow through the lives of generations . . . rivers flow from the past bringing memories and associations rich in meaning. And of course, rivers are also flowing into the future."[2]

The title of this book might seem incongruous, or even a contradiction wrapped in an oxymoron. *Rewilding*, *urban*, and *frontier* are not terms that enjoy a close association with each other. Rewilding is generally thought of as a massive effort to restore native species on a landscape scale. Restoration of wolves would be a prime example, but rewilding could even be as ambitious as bringing back mammoths or aurochs, the ancestors of wild cattle. Most humans on Earth now live in urban areas, from towns to megacities, but frontier? Besides being practically an anachronism, the term is so problematic that many historians refer to it as the f-word.

In his famous 1893 essay, "Significance of the Frontier in American History," Fredrick Jackson Turner declared the frontier experience as instrumental to the development of the "American character." While Jackson's "frontier thesis" has become an outmoded concept among academics and is fraught with overtones of racial, sexual, and ethnic violence, most Americans simply associate the frontier with the Wild West or Star Trek. Basically, a frontier is a place where the social mores are in flux and where laws and civil society have yet to reach, often leading to violence. Yet, freed of social constraints, a frontier can also be a place of freedom and reinvention where people can forge a new way of life and a new identity, even reshaping their reality. For Jackson, the frontier was also a process, a transformation from wilderness to civilization. Likewise, for this book, frontier denotes a meeting ground between past, present, and future perceptions of the natural world.

In its original meaning, frontier simply means an edge or border, as in the Spanish, *frontera*. In this book, the "urban frontier" refers to the border zone between domesticated and wild landscapes, as well as the ecological edge, or ecotone, between different ecosystems. Foregrounding the term frontier reminds us of the social and ecological violence that accompanied settler colonialism. Thinking of the frontier as a process, we can trace how the conquistador mindset that quickly "tamed" the frontier became institutionalized in U.S. bureaucracies, such as the Army Corps of Engineers with its emphasis on dams, channelization, and irrigation. Urban rivers illustrate how once something becomes established, innovation is stymied. Thus, a frontier can also be cutting edge, signifying a new approach or beginning. This is particularly apt in the sciences; in fact, the official journal of the Ecological Society of America is called *Frontiers in Ecology and the Environment.*"

For Jackson, the frontier was a moving target, primarily from east to west. While historically flawed (settlement of the American Southwest preceded much of the country), Americans encountered free-flowing rivers as they moved across and around the continent. Dependent upon reliable sources of water for everything from drinking to irrigation to electrical power, America's cities and towns boomed along watercourses: Pittsburgh, St. Louis, San Antonio, even Los Angeles and Phoenix owe their initial founding to a river. In addition, the frontier has long been a place where people reinvent themselves, experimenting with new forms of government, social mores, and personal, as well as collective, identities.

What then does rewilding the urban frontier in the Anthropocene mean? The chapters that follow explore that question from multiple perspectives. While the frontier suggests a place in transition, rewilding is the visionary process that moves us from a colonial and controlling mindset to adapting to multiple ideas and cultures and embracing an uncertain future. Rewilding is related to but distinct from restoration (see chapter 3). Restoration suggests a return to a past condition, whereas in rewilding, natural and human forces merge and create something new, and all of it happens in the most human of environments—the city.

The need for rewilding is a consequence of the Anthropocene, a concept arising in the late twentieth century. In 1989 Bill McKibben published *The End of Nature* in which he suggested that human activity, primarily the burning of fossil fuels, has so fundamentally changed Earth's climate that humans have replaced nature as the planet's dominant force. Eleven years later, Dutch chemist Paul Crutzen, whose pioneering atmostpheric research garnered a Nobel Prize, coined the term Anthropocene to signify a break from the geologic past. Crutzen argued that over the last two hundred years, human activity has so fundamentally changed the planet's climate, landscapes, nutrient cycles, and biotic systems as to mark a new geological epoch. Although in 2024 the International Union of Geological Sciences declined to formally designate the Anthropocene as a new geological epoch, the term remains salient in popular and academic use.

As a climate scientist, Crutzen was primarily concerned with ozone depletion and increased carbon dioxide levels—the first leading to increased ultraviolet radiation and the second to global warming. More recently, other scientists have weighed in, pointing out the extreme loss of biodiversity, with a quarter of all plants and animals on Earth in danger of extinction. Humans actively change the composition of ecosystems, not only through extinction and extraction, but we also reconfigure ecosystems through the introduction of non-native species, some of which become invasive and take over entire landscapes, such as kudzu in the American South. Equally significant, if not more, is the extent to which humans have replaced large mammals across the planet. Currently, the majority of the mammalian biomass on land consists of livestock (60 percent) with wild animals making up only 4 percent. Humans themselves compose 36 percent of the biomass of land mammals.[3]

Not only have we depleted and rearranged species, but we've also changed geochemical systems, synthetically producing nitrogen (from fossil fuels) and dumping this essential nutrient on crops worldwide, far beyond what they can absorb. The excess nitrogen spills into streams, rivers, lakes, and eventually the ocean, resulting in algae blooms that lead to extensive dead zones in aquatic and marine ecosystems. Throw in the increased phosphorous and other nutrients, in addition to the proliferation of pesticides, and our aquatic systems become a toxic brew. And, of course, we move massive quantities of earth around every day, leave asphalt behind us as if it were a snail trail, infuse nearly every biome with microplastics, and reconfigure and dam nearly every river on the planet.

The Anthropocene also marks a shift in our understanding. While much discussion about the Anthropocene laments (and for good reason) the recognition that humans have fundamentally changed the planet, the articulation of the term suggests a degree of self-awareness. This self-realization is also occurring on a social level, as suggested by the United Nations Declaration on the Rights of Indigenous People in 2007. Considering the long arc of human history, the twenty-first century marks a significant departure from the past. Ultimately, conservation in the Anthropocene will require a cultural and social shift, a noetic frontier if you will.

If in the Anthropocene the human footprint is ubiquitous, does that mean the end of nature? Have humans completely supplanted and replaced the natural world? If the COVID-19 pandemic taught us anything, it's that such a proposition is perilously arrogant. The realization that our own bodies host a dizzying array of microbes blurs the line between human and nature. Just as there are many ways of regarding the Anthropocene, there are multiple ways of considering the rivers that run through our cities and our lives, as these chapters suggest. How we grapple with our urban rivers reveals much about how we will approach the rest of the natural world in this new era. Will it be one of control and domination (repelling the frontier), or will it be one of uncertainty and adaptation?

For much of the nation's history, we vacillated between seeing nature as a natural resource for exploitation and seeing nature as sacred, to be preserved and revered. Restoration of urban rivers offers opportunities to reconnect people to the natural world, signaling a mending of the human-nature rift that has haunted us for the past two centuries. Might it be possible to integrate nature and society into a seamless whole so that both natural and human ecosystems can flourish and prosper?

Despite the popularity of national parks and wilderness areas, most Americans' interactions with nature are on or along the rivers that flow through our cities, towns, and suburbs. Urban rivers are where we recreate, whether it be fly-fishing on the Boise or kayaking on the South Platte. Urban riverways are where we walk, hike, bike, birdwatch, and commune with nature. But can such urban greenways really reconnect humans and nature, or are we creating a poor substitute for wildness? Are we simply trying to salvage what is left of a depauperate nature?

Regardless, urban rivers are receiving increased attention and money for cleanups, ecological improvement, and recreational development. Yet, this also produces conflict as a public with differing values negotiates priorities and access to public space, raising such questions as: Are urban rivers now primarily recreational amenities to enhance the new urbanism and economic development? In which case, how does gentrification and urban redevelopment (twenty-first-century analogs to nineteenth-century colonization and frontiers) along rivers affect the marginalized members of society?

Environmental narratives all too often skirt around issues of race and identity, both of which are integral to the work of restoration. Each chapter in this volume explores urban rewilding through a particular lens. As these chapters demonstrate, rewilding urban rivers coincides with a community's identity with its river. Urban rivers, once ignored, become reinvented. That process of identity formation is in flux and can yield a Chicago River of sewage and invasive species or a Boise River that becomes the city's pride or a Los Angeles River as shapeshifting as a Hollywood movie set.

Ecologically, urban rivers are essentially human creations, artifacts of industrial capitalism and human desires. How then do we "restore" such systems that remain morphologically and ecologically altered—dammed, channeled, and filled with exotic species? And what can urban rivers tell us about this new era we call the Anthropocene—a time when humans have fundamentally altered Earth's natural processes?

America's rivers have come a long way since the days of impoundment and pollution, and many rivers are experiencing various stages of restoration, renaturalizing, and enhancement. With the removal of more than 1,500 dams and the return of native fisheries, the nation's rivers are experiencing a revival. But in urban environments, simply restoring an ecosystem back to its previous condition is impossible as these rivers have been so modified. Nevertheless, some ecologists suggest that such ecosystems can play an important role in biodiversity conservation and landscape connectivity, especially in urban areas. Thus, urban rivers might be one of the best opportunities for rewilding in the Anthropocene—that is creating self-sustaining ecosystems capable of adapting to the rapid and cascading changes caused by human impacts.

Approaching urban rivers in this context challenges conventional notions of conservation and nature. Rather than seeing these degraded ecosystems as human systems designed for urban amenities or attempting to restore these ecosystems back to presettlement conditions—a Sisyphean task—we could instead choose to rewild these rivers, allowing them to chart their own evolution and resilience.

There are hundreds of rivers that could be included in this volume, and the reader might note the conspicuous absence of America's main stems (the Mississippi, Missouri, Columbia, Colorado, Potomac, Hudson, et cetera). The omission is intentional, for the focus on tributaries and smaller rivers allows us to highlight a particular issue or aspect of urban rewilding. In addition, often these smaller rivers are those we develop a closer relationship with and affinity toward. Rewilding urban rivers in the Anthropocene ultimately depends upon fostering a sense of place and developing a personal connection to our local waters. While it might be more apparent in some chapters than others, each of the contributors to this volume has a personal connection to their waterway. Whether it be cultural, recreational, or professional, each of the following chapters, contributed by an expert in their field, provides a unique perspective on the challenge of rewilding America's urban rivers.

Rather than attempt a full geographic scope of the United States, this book takes a thematic approach with each chapter exploring a particular aspect of the relationship between a city and its river while highlighting elements common to nearly every urban river such as flood control, pollution, and development. From issues of race and gentrification, to ecological restoration and public policy, to personal loss and reconnection, these themes surface throughout the book, indicating that for as unique as each river is, we face a similar suite of challenges in all of our communities. To paraphrase a great Russian novelist: all healthy rivers are alike; each unhealthy river is unhealthy in its own way.

Nonetheless, a master narrative emerges:

Most cities were built along rivers and for good reason, but this often meant contending with seasonal flooding, which could be catastrophic. America's creativity, engineering prowess, and industrial might led to attempts to control rivers and their floods through dams, channelization, and other engineering infrastructure throughout the twentieth century. Industrial development and inadequate sewage disposal led to water pollution and aquatic degradation. A significant shift occurred in 1972 when Congress passed the Clean Water Act to "restore and maintain the chemical, physical, and biological integrity of the Nation's waters" and gave the Environmental Protection Agency authority to regulate and eventually eliminate pollution and toxic discharges into navigable waters to achieve "fishable and swimmable" rivers and streams.[4] This remarkable piece of legislation provided the catalyst for a cascading effect on our urban rivers. Over the last half century, river restoration and cleanup efforts have led cities to regard rivers as amenities and economic assets with the corresponding increase in tourism and recreation. And yet, for all their successes, these river revitalization efforts are complicated by challenges of gentrification and increased recreational use.

Although the book is designed to be read sequentially, the chapters can also be read as stand-alone essays, depending on the reader's interests. Making for easy classroom adaptation, the book is divided into four parts, which can be read independently, and each ends with a potential path forward toward rewilding. Some readers may be captivated by the personal narrative of some authors, while others may be more interested in the pragmatic details of how cities have worked toward river restoration. Think of this as a mosaic, with each chapter, while whole unto itself, contributing to paint a larger complex image of how we can re-story our relationship to the natural world. By no means should this book be considered a Pollyanna affair, nor is it intended as a litany of environmental ills; instead, it is a presentation of possibilities, of multiple pathways forward.

Taken together these chapters illustrate the diversity of America's urban rivers and indicate that we have only just begun the process of rewilding. Engaging in this undertaking is not only transforming our waterscapes, but also ourselves and our relationship with the world around us. The process of rewilding itself, even if we don't quite agree on what this means, compels us to engage with nature in a fundamentally new way. The hubristic attempts to control and dominate the natural world have obviously failed as we've significantly altered the biological and geophysical systems on a planetary scale. But how do we make peace with an uncontrollable and unpredictable world, largely of our own making? Yet, as urban river rewilding demonstrates, this "brave new world that has such people in it" need not be a dystopian nightmare, but rather one in which humans actively engage with our surroundings in a way that recognizes and celebrates the more than human world to build resilient ecosystems and communities.

NOTES

1. Tara Lohan, "How Removing One Maine Dam 20 Years Ago Changed Everything," Revelator, February 11, 2019, https://therevelator.org/edwards-dam-removal/.
2. Quoted in Crane, "Setting the River Free," 144.
3. Baille and Zhang, "Space for Nature," 1051.
4. U.S. code 2018, title 33, chapter 26, section 1251.

BIBLIOGRAPHY

Baille, Jonathan, and Ya-Ping Zhang. "Space for Nature." *Science* 361, no. 6407 (September 14, 2018): 1051.

Crane, Jeff. "'Setting the River Free': The Removal of the Edwards Dam and the Restoration of the Kennebec River." *Water History* 1 (2009): 131–48.

Crutzen, P. J. "The 'Anthropocene.'" In *Earth System Science in the Anthropocene*, edited by E. Ehlers and T. Krafft, 13–18. Berlin: Springer, 2006.

McKibben, Bill. *The End of Nature*. New York: Random House, 1989.

Part 1 HEADWATERS

Reconciling Personal, Cultural, and Ecological Trauma

In many ways, part 1 serves as an overview for the book, presenting the broad sweep of the challenges and opportunities in rewilding urban rivers. Illustrative of America's urban rivers, the Spokane River embodies the master narrative of dams, degradation, pollution, and recovery. The Spokane is a prototypical river through a prototypical city. At 230,000 residents, Spokane is slightly smaller than the median U.S. city, but only half as dense, with a slightly higher per capita income, but spot on with the average age and poverty rate of other cities.[1] Unusual, but not unique to the Spokane River, is the significant role that Indigenous groups and other historically marginalized communities play in rewilding efforts.

The Spokane River is relatively short—only 110 miles—yet it drains much of the Idaho Panhandle and disperses heavy metals from a century of lead and silver mining along its route to the Columbia. Similar to many rivers, the Spokane watershed encompasses several land uses: rural, urban, and agricultural—all contributing to the river's degradation. And like most urban rivers, its structure, function, and species composition have changed radically since colonization. Dams and channelization transformed a meandering river into a series of slow-moving reservoirs. Industrial, agricultural, and municipal pollution have altered the river's nutrient levels and chemical composition, while exotic species, such as carp, bass, and walleye, have replaced salmon and steelhead.

In the first chapter, "Haunted," Heidi Lasher provides a personal metaphor for the conundrum we face when thinking of rewilding urban rivers.

Taking a canoe journey down the Spokane, she sees ghosts of the former river alongside the current reality. In the fast-paced Anthropocene, we bear witness to the changes that occur to a place within our own lifetimes. As Lasher demonstrates, far more than previous generations, we behold the past, present, and future of the natural world simultaneously through urban waterways. Rivers draw us in, as we seek physical contact, and provide a sense of grounding.

As Lasher describes the invisible pollution—the legacy of the frontier in residual heavy metals and toxic chemicals—beneath the bucolic surface of the river, she grapples with her own specter, noting, "The past does not condemn us, it only tethers us to the future." One of the characteristics of urban rivers is that they define a place and engender a deep personal connection to that place, whether it be individual or multigenerational. In exploring the imperfect river that is, Lasher embraces an unknowable future, thus underscoring the difficulty of conservation in the Anthropocene.

Embracing a much deeper past that also informs an uncertain future, Margo Hill's essay, "We Buried Our People along the River," explores her tribe's spiritual connection to their namesake river and the tragic history of displacement at the hands of white settlers. Equally poignant was the relocation of burial sites during the construction of the Grand Coulee Dam during the 1930s. While the burial relocation was bungled, Hill reminds us of the resilience of both the river and Native peoples. Despite the continued presence of massive dams that block anadromous fish passage of the entire upper Columbia, the Spokane Tribe and others have begun to reintroduce salmon into the Spokane River. Taken together, these two chapters illustrate how urban rivers are products of what is upstream (mining and heavy metals) and what lies downstream (dams and graves).

The last chapter, "Terms and Conditions Apply," frames the rest of the book, as it examines a pond on a college campus to illustrate how changing conceptual ideas about nature manifest themselves on the landscape. Such changes demonstrate how ecologists are beginning to think about hybrid and novel ecosystems that comprise urban watersheds and how we can approach conservation in the twenty-first century. Challenging our traditional notions of conservation as protecting nature from humans, I suggest we consider human and natural habitats as one in the same, and we make resilience our conservation goal in the Anthropocene.

Sometimes a single event has the capacity to reconcile historical, cultural, and ecological trauma. One warm August afternoon in 2022, several hundred people gathered at the confluence of Hangman Creek and the Spokane River, a place currently called People's Park, so named for the free camping during the 1974 World's Fair. The confluence, a wide sandy expanse of ponderosa pine and serviceberry and riddled with social trails, however, has a much deeper human history, going back some twelve thousand years to the last ice age. Depending how you parse the meaning of "settlement," you could argue that this is one of the oldest continually inhabited sites in North America. On the heels of retreating ice, the Salish people established a fishing camp here to harvest some of the millions of Columbia River salmon that returned to their natal waters on the Spokane and its tributaries. Over millennia, these fishing sites Spokane Falls, this confluence, the one downstream at the Little Spokane, and Kettle Falls on the Columbia—became gathering places for peoples from across the Columbia Plateau to fish, trade, socialize, play, and worship. Today, only a bend in the river obscures the high-rise buildings, banks, offices, department stores, and hospitals from the park.

More than one hundred years after the last salmon bumped its nose against the falls, Native and non-Native alike gathered again to witness salmon thrashing through the late summer shallows. Unlike the initial salmon release on the Little Spokane that Margo Hill describes, this event was open to the public, and school buses dropped off kids from across the city, county, and nearby reservations. The city bus stop at the parking lot cast off a steady stream of onlookers, cars lined the roadway, and despite the midweek heat, people continued to converge at the park providing a festive atmosphere to the occasion.

The ceremony opened with brief speeches from several tribal elders. Unlike a white or Western culture that might highlight individual efforts and accolades, the elders focused on the river and the salmon. Spokane tribal chairwoman Carol Evans thanked the river itself for bringing back the salmon, saying, "Thank you, River, for all you do for us." Tribal elder

Warren Seyler noted how restoring the salmon helps the tribe recover from historical trauma. From food sovereignty to cultural esteem, salmon restoration is a living embodiment of environmental justice.

Following the ceremony, tribal and community members formed a twin human chain leading from the tanker trucks parked high on the bank down to the river. A pair of fisheries biologists netted the adult salmon from the tank, placed them in a rubber tube, and passed them down the line. At the river, tribal biologists lovingly set the rubber tube into the water. They pulled each fish out by the tail and swished it back and forth to get the cold, oxygenated water moving through its gills and then slowly let go. With the clear water of late summer and the circling osprey above, the salmon zipped toward the deeper pools and hid under rock ledges. One older gentleman watching from the high bank remarked, "I never thought I'd see the Berlin Wall come down, but that's nothing compared to this."

These Chinook salmon were raised in a hatchery on the Columbia and then flushed and barged past multiple dams and out to sea. The salmon migrated to the Aleutian Islands, growing fat on the ocean bounty and returned up the Columbia where they were collected at Wells Dam, near Lake Chelan. The tribal biologists then trucked them to the release site on the Spokane River so that they might lay eggs and die. Their spawn, should any survive, have little chance of making it to the ocean. With waters backed up by fourteen dams, most the way downriver is one reservoir after another, where the slack water confuses the migrating salmon and exposes them to predation by birds and larger fish. For adult salmon, the return voyage is even more daunting. In addition to negotiating the nine dams on the Columbia with fish ladders, the last two—the massive Chief Joseph and Grand Coulee Dams—have no fish passage. And then there are three more on the lower Spokane River, also built without fish passage. Even if the biologists truck salmon upstream around the dams, and if the fish reproduce in the tributaries of the Spokane River, the young fry would still be functionally landlocked behind multiple dams and reservoirs.

Despite these odds, in 2017 the Spokane Tribe released 752 juvenile Chinook salmon in Tshimikain Creek, a tributary of the Spokane River. Outfitted with Passive Integrated Transponder tags so they could be detected as they passed by the dams, somehow 75 fish traveled 680 miles to the mouth of the Columbia, past three dams without fish ladders, past nine

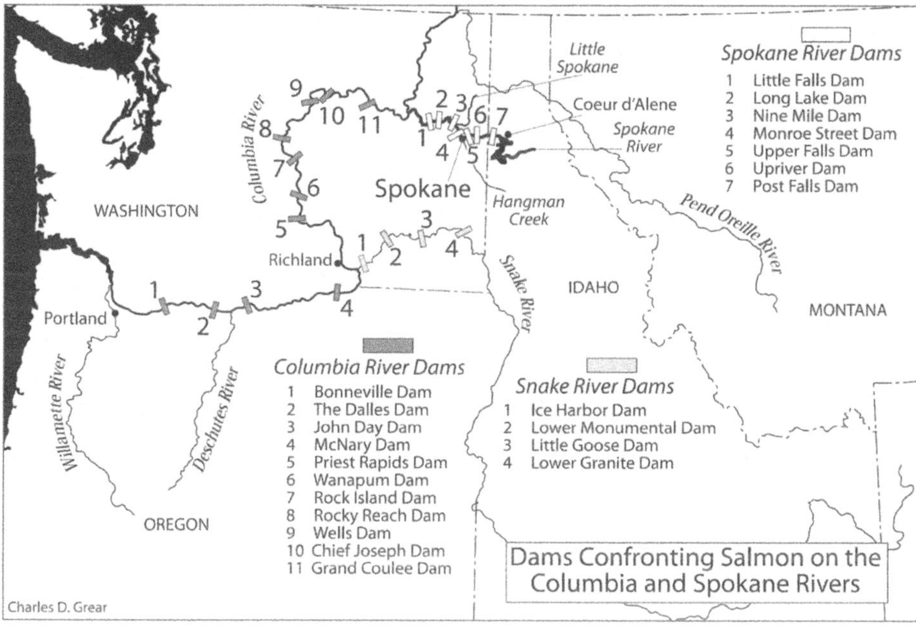

1. Map of dams on the Columbia, Spokane, and Snake Rivers. Created by Charles David Grear, used by permission.

more, past the reservoirs, past cormorants and predacious fish, and out to sea. Miraculously, two years later, a lone female adult Chinook from Tshimikain Creek made the return journey all the way back to Chief Joseph Dam, swimming upstream as far as she could. Although policy constraints prevented the tribe from bringing her back alive, her remains were returned to Tshimikain Creek. Dubbed Nucucšnetkw, "she who retraces her steps," she symbolizes the possibilities of rewilding.[2]

Even more surprising was the return of Sye'us'uslsh, "she who repeatedly dives," in 2022. Two years earlier, near the headwaters of Hangman Creek, sixty miles upstream from Spokane, the Coeur d'Alene Tribe released 1,500 juvenile Chinook salmon. This time two fish managed to get past five dams (three on the Spokane River, plus the two on the Columbia) without fish passage, out to the ocean, and back up again to Chief Joseph Dam. There Sye'us'uslsh was captured and transported back to Hangman Creek, where she was released alive just above one of the city's golf courses, a short distance from downtown Spokane.[3]

These initial releases demonstrated that salmon restoration to the upper Columbia is theoretically possible, so the Upper Columbia United Tribes (UCUT) developed a twenty-one-year feasibility study, called phase two, with a price tag of $208 million, far beyond the capacity of UCUT. But then, in September 2023, the Biden administration, acknowledging the federal government's culpability in destroying the salmon runs of the upper Columbia Basin, reached a historic agreement with the tribes to provide $200 million from the Bonneville Power Administration over twenty years to implement phase two.[4]

Nevertheless, none of the dams are coming down anytime soon as they provide the majority of the region's electricity. Therefore, many observers considered the releases mostly symbolic. But doing so would miss the slipperier point: rewilding requires imagination. It demands we think beyond the practical and pragmatic and recognize the resilience of nature. Rewilding will span many generations and the steps might not always be linear. Seeing those big fish thrash about and move up and down the river in which they belong—as well as letting the kids from the reservations of Spokane, Coeur d'Alene, and Colville carry the inner tubes and feel the weight of the salmon before passing them on down the line—is about witnessing and imagining possibilities. On August 24, 2022, 141 adult Chinook salmon swam in the Spokane River for the first time in 112 years. A few days later, biologists found thirty-two redds (rock nests built by female salmon to protect their eggs) and observed the salmon spawning between the frisbee golf course and downtown Spokane.[5] That is rewilding. That is how we tackle the Anthropocene.

NOTES

1. "Spokane, Washington Population 2023," World Population Review, accessed October 1, 2022, https://worldpopulationreview.com/us-cities/spokane-wa-population.
2. "She Who Retraces Her Steps," Spokane Tribal Fisheries, accessed April 30, 2023, https://spokanetribalfisheries.com/news/.
3. K. C. Mehaffey, "Against the Odds, Sye'us'uslsh Comes Home," Spokane Tribal Fisheries, July 15, 2022, https://spokanetribalfisheries.com/news/. The fate of her sibling is unknown.
4. "The Phase 2 Implementation Plan: Testing Feasibility of Reintroduced Salmon in the Upper Columbia River Basin," Upper Columbia United Tribes, August 4,

2022, https://ucut.org/water/phase-2-implementation-plan-testing-feasibility-of-reintroduced-salmon/; "Biden-Harris Administration, Tribes Reach Historic Agreement Supporting Efforts to Restore Healthy and Abundant Salmon Populations to Upper Columbia River Basin," U.S. Department of Interior, press release, September 21, 2023.

5. Conor Giorgi, "Preparing for Salmon Homecoming" (presentation at Spokane River Forum Conference, Spokane WA, April 26, 2023).

1 Haunted

Heidi Lasher

The Spokane River slinks away from the northern tip of Coeur d'Alene Lake like an introverted guest at a party. If you stood on the shore, you would see no perceptible difference between the lake and the outlet to the river: it's all flat and still. But if you looked closely, you might see how the lake pulls to the west, becoming a bay that squeezes into a lane the size of a suburban street. Disguised as such, the river paves a smooth, liquid road for thirteen miles, reaching a cul-de-sac at Post Falls Dam. Only once it is past the dam does the Spokane assume the likeness of a river.

I canoed this upper section of the river one recent spring with my friend Marit, who grew up in Spokane and had recently returned to the city with her husband and daughter. A former marketing executive, Marit now works as a spiritual-regression therapist, which I believe makes her a kind of ghostbuster. In her words, she helps people examine their past lives. She tells me that her clients seek her out because they're desperate to untether themselves from troubling emotional trauma—trauma that extends beyond their lifetime—and they cannot do it alone. They need someone to guide them.

I am baffled by the number of people who seek her services. I do not believe in past lives or spirits or ghosts. I find navigating my current lifetime complicated enough without dredging up an imagined past. We find plenty of other things to discuss. When we launched our canoe on the upper Spokane River that day, we talked about the twin plagues of goose poop (there was a lot of it on the beach) and regional growth.

The north bank of the river (the bank to our right as we paddled downstream) had experienced a recent surge in commercial and residential development. In 2018 the city of Coeur d'Alene purchased forty-seven acres of waterfront from an old lumber mill and sold the acres to developers to build restaurants, a marina, a public park, and a row of new waterfront houses (which, on that day, were in various stages of construction). Despite zoning laws that require a twenty-five-foot setback to maintain riparian habitat on the riverbank or lakeshore, this development, like many others, appeared to be exempt from the law. I learned later that the twenty-five-foot setback includes vertical feet, which means you can build a twenty-five-foot-high retaining wall between your home and the river and remain in compliance with the law. Each new house we passed featured a tidy green lawn shored up by concrete bulwarks and a dock.

The south bank of the river, which we passed on our left, looked completely different. Perhaps because it's harder to access—requiring a long drive downriver to a bridge and a long drive back upriver on a winding road—the left bank appeared to be set in a different time period. On this bank we passed ramshackle cabins with big front porches and rocking chairs. Overgrown tea roses and gnarled cherry trees sat back from the bank where old rowboats hid in the tall grass.

We paddled quietly down the middle of this rift in time, looking left toward the past and right toward the future. It gave me an eerie feeling to wander so freely in time, between one set of ideals and another. The two sides of the river seemed at odds with each other, and as I paddled between them, they tugged at two sides of myself: the optimist and the cynic.

Before that year I hadn't understood much about the Spokane River. I didn't have a good sense for it—where it came from, where it went, or what it passed along the way. I'd hardly glanced at it.

Something had changed in the preceding months. For no discernable reason, the river became visible to me. It drew me to it like a magnet. I began running and biking alongside it, stopping awkwardly to snap pictures of it. When I wasn't near the river, I scrolled through pictures I had taken, read books about the river, and studied maps to get a sense for its size and scale. Before long, I craved physical contact with the river. I yearned for a visceral experience of the cold and heavy spring flood, to feel it slice through my arteries and rush past my organs on its way down to my feet and out again. I could not explain this urge. It felt instinctual and compulsive. When I admitted my obsession to my husband or friends, I felt jittery and exposed, and I'd almost giggle the river's name out loud, aware of how absurd I sounded.

My newfound passion for the river confused people who knew me well, for I was not known to exhibit zealous behavior. My astrological sign is Libra, represented by the scales, eternally balanced and even-keeled. I am part stoic, hardworking Norwegian, part pragmatic, humorless Swiss-German. I majored in economics. I don't fall for much.

Plus, the Spokane River is relatively unremarkable. It is neither pristine, nor particularly wild. It's not very long, not known for its whitewater or fishing (though it offers both), not easy to access. There are healthier and more productive rivers nearby in Idaho and Montana. After it departs from Coeur d'Alene Lake in Idaho, the Spokane slides past uninspiring homes, trailer parks, ponderosa forests, and undeveloped land. Past strip malls, gravel pits, and industrial parks. It pours straight through the downtown core of the midsize city of Spokane, meanders through a state park, then along the Spokane Indian Reservation, and finally into the Columbia River. There are seven dams along the Spokane River, leaving a good portion of the water stagnant and choked with noxious weeds and algae. The free-flowing sections can be littered with garbage: empty beer cans, carpet remnants, tires, and crumbling concrete. In flat-water stretches, the river bank is hardscaped and lined with docks and motorboats. The downtown section of the river is festooned with placards warning residents to "Stay Out, Stay Alive." This is to prevent unsuspecting swimmers and boaters from being sucked into the hydraulics of the dam or being dragged over the Spokane Falls. But people tend to interpret the message generally and steer clear of the river as a whole.[1]

As for me, I had begun to paddle the entire length of the river. By the time I canoed with Marit, I had already traveled up tributaries, sat in on meetings, and studied historical pictures of the river when it was criss-crossed with railroad tracks. I had learned the names of birds and the locations of beaver lodges and the natural history of the signal crayfish, a spectacular orange and turquoise mini-lobster that hunts by the light of the moon. The river had appeared at a turning point in my life. I had become disenchanted with work and decided to spend more time at home. My kids needed less of me, but I seemed to require more of myself. I'd lost track of who I was. As I wandered the house, wondering what I'd do next, I began to look around—not just metaphorically, but literally—where on Earth was I? What were my geographic, spatial, and temporal coordinates? Where did I come from, and where was I going?

The Spokane River emerges from a lake. Considered locally to be one of the gems of North Idaho, Coeur d'Alene Lake is a large and natural lake, kept unnaturally full in summer by the Post Falls Dam. There are summer camps along the lake, ramshackle cabins, and monstrous second homes. In summer the lake buzzes with motorboats: puttering pontoon boats, fishing boats, waterski boats, wave runners, and the occasional blast of a jet boat. Despite recent development, the water is still bordered by coniferous forests that rise toward a clear blue sky. Because of all the intricate folds of land around it, the lake is pleated into 135 miles of shoreline. It's a long body of water from north to south. A boat can travel south for 25 miles from Coeur d'Alene and not quite reach the other end.

At the south end of the lake is Indian Country. French fur traders who thought highly of the tribe's aptitude for trading named the Salish-speaking people of the region, "Coeur d'Alene" (Heart of the Awl). The Coeur d'Alene Tribe once occupied about five million acres of land surrounding the lake, where the clear water, thick forests, and free-moving rivers provided a cultural and economic mainstay. When white traders and speculators arrived in the area in the early 1800s, the tribe tried several times to negotiate a treaty that would keep the lake in tribal hands. But the productive forests, ample hunting, and the promise of silver in the Idaho Panhandle, which was discovered in 1863, made the lake an intoxicating prize. Eventually, the U.S. government offered the tribe a paltry 345,000-acre reservation, which included only the southern third of the lake. The northern two-thirds of the lake, its northeastern tributaries, and northwestern outlet would be exploited and remade by the newcomers.

For the next one hundred years and more, the aptly named Silver Valley, east of and upstream from Coeur d'Alene, would become the country's largest and most productive silver mining operation, also producing lead, zinc, gold, copper, cadmium, antimony, cobalt, and uranium. A smelting operation in the valley would send plumes of poison into the air, which would fall as acid rain. Trees on both sides of the valley would die, creating a reverse oasis, a barren desert inside a green forest, with a toxic river running through it.[2]

The South Fork of the Coeur d'Alene River—one of the two major tributaries to the lake—conveyed mine tailings downstream. The metals tumbled over rocks, falling like dust onto the riverbed, swirling in eddies, and lifting and rolling, settling on top of wetlands and landing finally at the bottom of Coeur d'Alene Lake. Today, a thick slurry of heavy metals—some seventy-five million tons—covers the lake bottom. One must be careful not to stir it. Signage all along the lake warns visitors not to stray off the paved trails or dig in the sand or dirt along the lake unless the soil has been remediated, but some of us are better than others at reading signs. Migrating tundra swans, who dip their long necks in the sediment to feed, inadvertently poison themselves by the hundreds every April and May.

You wouldn't know any of this if you were paddling with us on the outlet to the Spokane River. You wouldn't necessarily believe it even if you owned property on the lake. That the lake has become a dangerous repository of heavy metals is not a big secret, but our brains have a hard time believing in things we cannot see. From the surface, Coeur d'Alene Lake looks like any other lake in Idaho: sparkly, fresh, and swimmable. It looks like the kind of place you would want to bring the family. It looks like a fabulous real estate investment. And yet, as Marit and I paddled along the river disguised as a lake, something felt unsettled, haunted even.

We passed a guy on the right bank, mowing his lawn. It was a nice lawn, free of weeds, well-trimmed. Michael Pollan, who wrote an essay about our country's tyrannical obsession over lawns, would have called it an American lawn. The carpet of grass made me think of phosphorus, an ingredient in lawn fertilizer. In Greek, it means "bringer of light." Its Latin equivalent is "lucifer." Phosphorus is a chemical element, invisible, and glows faintly green when exposed to oxygen. It is abundantly available in soil and is a common ingredient in many household items like pesticides, detergents, and (as mentioned) fertilizers. And it can wreak havoc on freshwater ecosystems.

Phosphorus levels in Coeur d'Alene Lake have doubled since the 1990s, trickling in from upstream logging and agricultural operations, leaching out of septic tanks and fertilized lawns, and seeping in from soils disturbed by new development.[3] I wondered if the guy mowing his lawn knew this. And if he did, whether it bothered him.

Lakes that absorb phosphorus tend to grow floating rafts of algae, which can blanket the lake in suffocating muck. Algae are not just a nuisance for swimmers and jet skiers, they also block sunlight from reaching the lake floor, which makes it impossible for native underwater plants to photosynthesize and make oxygen. Algae further deprive the lake bottom of oxygen when they die and sink to the bottom because the bacteria that decompose the algae use available oxygen to do so.

Scientists from the Environmental Protection Agency and Coeur d'Alene Tribe have documented the level of dissolved oxygen in the lake since 1991 and have seen it steadily drop, particularly in the more shallow, southern portion of the lake. Most of the lake is currently somewhere in the range of six to eight milligrams per liter. When dissolved oxygen drops below about two milligrams per liter, water can become hypoxic, causing organisms living in the water to suffocate. Eventually, water can become anoxic, devoid of oxygen, which is actually happening in late summer in the southern end of Coeur d'Alene Lake.[4]

Oxygen, so essential to aquatic life, plays a critical secondary role in the lake: it traps the heavy metals sitting on the lake floor. Like a gaseous lid, oxygen keeps the lead, cadmium, and zinc from floating up and circulating in the water column. When, and if, oxygen levels drop low enough, the lid will dissipate and the metals will rise, zombie-like, from the lake bottom, slowly poisoning everything in the lake and the river below: plankton, fish, birds, humans, and all.

On the day of my paddle with Marit, it was hard to imagine any of this happening. It seemed far-fetched to believe that one person's building project or lawn fertilizer could change the chemistry of a huge, shimmering lake surrounded by woods in the middle of Idaho. But I remember looking up at the sky from the unnaturally placid river and feeling a humming tension between the two sides of the river, between the past and the future, wondering how to move forward.

I may have been projecting my own sense of paralysis. In the past, I felt like anything was possible; time seemed endless, and no goals felt out of reach. But I had recently lost my momentum. I had lost touch with my ambition. No longer young or full of energy, I could not envision my next steps. The daily news documented a world in crisis. The years ahead filled me with dread. Night after night, I would bolt awake at 3:00 a.m. in a pool of sweat with my heart racing, impotent to stop or slow the frantic pace of human destruction and consumption, the changing climate, political division. I wanted to go back in time, back to some kind of beginning. I felt as if I were pulling with my whole body, mutely screaming for everyone to stop. Just stop.

There was a time when the Spokane River was full of clean, clear water that tumbled over stones, alive with native trout, migrating salmon, insects, ducks, and otters. The water would have been shaded by cattails, willows, and pine trees. It would have been peaceful and quiet, or perhaps loud with noise of its own making. I wanted to go back to that Eden.

If we went back just 150 years, I reasoned, we could refill the mines, unbuild the dams, save the salmon, protect the trees, safeguard the water. If we could unwind the clock two centuries, we could roll back our dependence on fossil fuels, reverse climate change. White settlers could start fresh with Native Americans and undo smallpox and land grabs, boarding schools and reservations. Going back a little further, merchants could reverse the slave trade, bishops could rethink the Crusades, warriors could subdue the conquering impulse.

"You're stuck in the past. See how your lower back is rounded?" My sister, a yoga teacher, had come for a visit that spring. She shuffled behind me and pushed both feet against my back, causing me to arch slightly and sit forward on my perineum. "There," she said. "You can sit longer in this position. You are more open and balanced. You can face the future this way."

My physical therapist said something similar when I went to see her about sciatic nerve pain. She watched me for a minute after pressing her palm into my glutes. "You're standing on the back of your heels," she said.

"Try shifting your weight forward." After a week of conscious effort to move my weight toward the balls of my feet, the pain disappeared.

It was time to let go of the past, not to ignore it. The trauma we've inflicted on each other and on the environment is real and durable. It is embedded in the landscape, it sits on the bottom of the lake, and it hovers invisibly over tense meetings between the tribe and the state. To ignore it is to live in a perpetual state of denial. But to recreate it is folly.

In 2002, after cleaning up the worst of the contamination in the Silver Valley and bringing lead poisoning levels down to an acceptable level, the U.S. Environmental Protection Agency turned its attention toward cleaning up contamination in the larger Coeur d'Alene basin. But local and state political leaders, wary of the stigma of being labeled a Superfund site, balked at the proposal.[5] Declining Superfund designation, they cut themselves out of federal funding and federal leadership, leaving the fate of the lake in the hands of old adversaries: the state on one end and the tribe on the other.

The Coeur d'Alene Tribe and the State of Idaho have never quite agreed on the lake's plight, much less its recovery. As the oxygen level continues to drop, the question of how to move forward continues to be averted and delayed. In 2020 the governor of Idaho asked the National Academy of Sciences to take over discussions on the future of Coeur d'Alene Lake and review the data that has been collected since discussions began on remediating the damage to the lake. By the time they finish, another eighteen months will have passed.

Eighteen months is not a long time in the scheme of things. It's a short delay after two decades of active cleanup on the South Fork of the Coeur d'Alene River. It's a fraction of the time that's passed since the lake bottom was covered with heavy metals. It doesn't register as time at all on the geologic scale. And yet, every minute that passes without action brings the lake that much closer to an ominous and potentially hostile future. The kind of future that arises when we refuse to acknowledge where we've been and what we've done.

On the river, Marit and I struggled against an afternoon headwind. Rounding a bend, we encountered a mansion on the right bank so out of place in the riparian landscape that we stopped paddling to gape at it. Ornate with columns and sculptural flourishes and exotic, twisting blossoms, the house contains twenty-eight thousand square feet of living space with thirteen bedrooms and thirteen bathrooms and a ten-car garage and two boat slips. It had been built by a founder of Amway, the cleaning-supply pyramid scheme. It looked as if it had been plunked on the land like a Southern belle who'd fallen for the wrong guy, followed him to Idaho, and now awaited her rescue. A sale sign asked $8.5 million, down from $16 million, in a part of the country where the median home price is $332,000.

To our left, our ears picked up the mournful note of a tenor saxophone. Tearing our eyes off the mansion, we found on the opposite shore a man with a grey beard, wearing a black cowboy hat and sitting on an old cantilevered wooden deck in front of an old cabin. He was pouring his soul into a brass saxophone, which had been hooked into an amplifier that itself was plugged into a long extension cord that draped over the decking and disappeared inside the cabin. The wind began to howl and carry away the mournful notes.

I felt, in that moment, as if my choice were to either regret a bygone past or to embrace a misguided future. Instinctively, Marit and I dug our paddles forward. Against the stiff wind, we immersed ourselves in the task of making progress, propelling ourselves along a river that could no longer move.

I thought, then, about Marit's job of releasing people from trauma experienced in past lives. I wondered if she could release us from the trauma of our collective past lives. Could she help us let it go? Would letting go help us face the future? Help us change our ways?

I didn't understand it then, but I realize now that the crux of my anxiety over the river and the central question of my midlife emerged from a struggle to manage change. I had been wishing to go back in time, to reverse environmental damage and restore the lake and river to an arbitrary Eden. I had been looking longingly over my shoulder at my younger years when the horizon looked longer, the possibilities endless. I had been begging for time to stop.

I'd forgotten that Earth and everything on it is in a constant state of flux. Grass grows; tadpoles become frogs; generations pass; species go extinct. Even mountains move. Glaciers carve new landscapes, rivers change course, volcanoes erupt, and winds erase and pile new contours. Everything is capable of—designed for—change. The past does not condemn us, it only tethers us to the future. It illuminates our choices. Although my generation did not run the mines that polluted this lake, we have an opportunity to mitigate the damage. We did not build the dams, but we can choose how long to keep them operating and at what cost. We did not negotiate treaties between Native Americans and the U.S. government, but we can uphold what was promised and acknowledge what was lost. If we cannot restore the lake to what it once was, perhaps we can imagine what it might still become.

Maybe the second half of life could be as nourishing as the first.

Marit and I finished our paddle in silence and drove together back to Coeur d'Alene. When she dropped me off at my truck, I walked to the lake edge, where the river pulls the water away to the west, and kicked a pile of goose poop. Since the beginning, this unremarkable river had been moving toward something bigger than itself, relentlessly seeking its own transformation. As I watched its quiet movement, I felt my weight shift forward.

NOTES

1. Emily Benson, "A Dangerous Cocktail Threatens the Gem of North Idaho," *High Country News*, June 24, 2019.
2. Snow, *Living with Lead.*
3. Molly S. Wood and Michael A. Beckwith, *Coeur d'Alene Lake, Idaho: Insights Gained from Limnological Studies of 1991–92 and 2004–06*, U.S. Geological Survey, 2008, 16; Craig Cooper, "Overview of the Phosphorus Inventory Report" (presentation to the Committee on the Future of Water Quality in Coeur d'Alene Lake, National Academy of Sciences, Engineering and Medicine, Washington DC, May 4, 2021).
4. *Optimization Review Report: Remedial Process Optimization Study*, Kootenai County ID, June 30, 2020, https://semspub.epa.gov/work/10/100250521.pdf.
5. National Research Council, *Superfund and Mining Megasites*, 4.

BIBLIOGRAPHY

National Research Council. *Superfund and Mining Megasites: Lessons from the Coeur d'Alene River Basin.* Washington DC: National Academies Press, 2005.

Snow, Bradley. *Living with Lead: An Environmental History of Idaho's Coeur d'Alenes, 1885–2011.* Pittsburgh PA: University of Pittsburgh Press, 2017.

2 We Buried Our People along the River

Margo Hill

The Spokane tribal people have always been and always will be physically, culturally, and spiritually connected to the river and its salmon. From the time I was young, I can remember gathering along the Spokane River in ceremony and song. We called both the city and the Spokane River Sƛ̓x̣etk^{w}, which means “Place of Fast-Moving Waters.” Even though I grew up on the Spokane Indian Reservation, I did not truly appreciate the sacred power of the river until I understood the story of how the Spokane people came to be.

In 1907 Chief Oliver Lot recounted our creation story to Indian agent Maj. R. D. Gwydir:

> A long time ago the land around where Spokane Falls [the former name of the city] are now, and for many days’ journey east of it, was a large and beautiful lake. . . . On its shores were many villages with many people. The Indians were well fed and happy, for there were plenty of fish and elk. But one summer morning the people were startled by a rumbling and a shaking of the earth. The lake waters rose. Soon the waves became mountains of water that broke with fury against the shore.
>
> Then the sun was blotted out, and darkness covered the land and the water. Terrified, the people ran to the hills to get away from the pounding water. For two days the earth rumbled and quaked. Then a rain of ashes began to fall. It fell for several weeks.

At last, the ashes stopped falling, the waters of the lake became quiet, and the Indians came down from the hills. But soon the lake began to disappear. Dry land rose where the water had been. Many people died, for there was nothing to eat. The game animals had run away when the people fled to the hills, and no one dared go out on the lake to fish.

Some of the water was flowing westward from the lake that remained. The people followed it until they came to a waterfall. Soon they saw salmon coming up the new river from the big river west of them. So, they built a village beside the waterfall in the new river and made it their home.[1]

Compare our Native creation story to the currently accepted scientific version of events known as the ice age floods:

> Between 18,000 and 13,000 years ago, glaciers advanced and retreated over northern Washington. This ice formed a fragile dam on the Clark Fork River, which slowly filled with water, creating Glacial Lake Missoula. The lake spanned hundreds of miles over present-day Montana. Periodically, the dam burst open, sending the entire lake hurtling down the western slope of the continent. It is estimated that these floods carried as much as fifty cubic miles of dirt and rock with them. Over time, the ice dams reformed, creating a millennia-long flood cycle that might have taken place 60 or even 100 times.[2]

These lakes receded, and the Spaqni (anglicized as Spokane) made the banks of the Spokane River their home for the next ten thousand years. Geologists required decades of field work to discover what the local Native Americans had known for generations. These repeated cycles described by the scientists created the homelands of the Spaqni and form the foundation of the creation story told by my people. Our tribal historian, Warren Seyler, believes that scientists heard Chief Lot's version of geographic land formation and then found the scientific evidence in the mid-twentieth century to back up this oral history.[3] Regardless, these ice age floods corroborate Chief Lot's telling of the creation story. Although Lot passed away in 1909, he helped preserve our tribal oral history and thereby our organic connection to the land and water by recounting our creation story for generations.

Lot also helped secure the Spokane Indian Reservation in 1881. His Indian name translates to "Walk on Striped Face," as recalled by tribal elders Robert Sherwood and Vi Seymour. According to Seyler, another name Lot was known by was Whis-le-poo-sum, or "Walking through the Reeds." In recognition of his contribution to tribal society, Sadie Boyd, my *tupiye* (great-grandmother) and a notable historian who lived to be 103 years old, would later erect a monument to Chief Lot.

At the dedication of the monument to Chief Lot on October 23, 1969, Boyd said she was afraid we would fail to remember and appreciate him. My great-grandmother spoke to the assembled crowd at the Presbyterian church on that day, stating, "Chief Lot was a leader who achieved much for us."[4] She spoke about how he had traveled to Washington and fought for us to have our reservation and a school. Boyd also spoke about the agreement with the "whiteskins" so that the Spaqni people might live in "contentment." Speaking in Salish, Boyd then told her own version of our creation story—how we, the Spaqni, came into existence. Boyd ended stating, "šey̓ huł smłič u xʷlíʔil. u kʷeṅt tl̓ še čsel u x̣cic łuʔ snxʷméneʔi, łuʔ sntuʔtʔúlixʷi, u łuʔ scqescíłni." In English: "They lived from all those salmon. And from those two people arose . . . the Middle Spokane, the Upper Spokane, . . . and the Lower Spokane," the three bands of the Spokane people. Today the Spokane Tribal Language House uses Boyd's version of the creation story to teach young and old tribal members the language.

What does it mean to teach and learn the creation story? It tells of how we come from the river and how we are connected to the river. Dave Archambault Sr., a Lakota tribal educator, stated that the right to teach our Native children our values, beliefs, language, and customs begins with "knowledge and understanding of the creation story—it is essential to impart this knowledge of how we came to be."[5] Long ago our ancestors taught our young people science, morals, survival, and critical thinking through the stories of our animal people. We use animals who think and talk like people as characters in our stories. Coyote, one of our favorites, makes all manner of mischief, and we learn from his mistakes. My great-grandmother told me the stories of the animal people who lived along the river and how our people would survive if we all worked and shared our salmon.

The Spaqni have always been connected to the water of the river. One notable time of year that we gathered on the banks of the river was for the

Easter sunrise service. Although a Christian holiday, we recognized it by meeting on our riverbank, a sacred location, to honor our ancestors. As a kid, I was not quite sure why our congregation from the Presbyterian church would gather at the Cornelius Cemetery, on the Spokane Reservation, to pray and sing songs. We did it (religiously) at sunrise. I did not understand why we always went to the cemetery in West End that overlooked the river. But our elders insisted on it, and my mother got us all up at the break of dawn to be there. As I grew up, the purpose of the yearly rite became clear. We went to the river to pay homage to our ancestors who were buried along the river. Our ancestors and the river gave us our way of life. While my great-grandmother always prayed in her Salish dialect, not all of us still spoke the language. Nevertheless, as we stood in the morning cold, though we sang songs and said prayers in English, our Indian hearts connected in our language. These riverbanks held not only the stories of how our people lived, but also the traumatic memories of how we were treated by the U.S. government.

For example, I learned the story of Chief Spokane Garry's historic bell from my tupiye. Sadie was married to Thomas Garry, who was the nephew raised as a son by Spokane Garry (1811–92). Back in 1830, Chief Garry built a traditional lodge church from tule-reed mats at Drumheller Spring on the north side of the river, right in the middle of the city of Spokane. The area housed an Indian encampment, and Spokane Garry ran a school there. Later, however, the encroachment of white settlers forced Spokane Garry to relocate west of town.[6]

My great-grandmother eventually found the bell in a pile of junk, had it welded because it had cracked, then in 1972 placed it, fully refurbished, atop a seven-foot granite monument in the Chief Lot Cemetery near the Presbyterian church in Wellpinit on the Spokane Reservation. She had the story of the bell engraved directly onto the monument:

> Thomas Garry, the adopted son of Chief Spokane Garry, took care of his father's bell. Thomas told Sadie, his wife, the history of the bell—how his father received the bell from the original First Presbyterian Church and took it to his cabin on Pleasant Prairie. A whiteman stole his farm. Spokane Garry rescued his bell taking it by horse back to Indian Prairie where he was helped by three friends: William Three-

Mountain, Solomon Scott, and Thomas's brother Tidas Williams, to set up a new church for Christ and the bell. After three years they were forced by whitemen to move again, this time to Spokane [Indian] River Church. From the rising waters of the [Grand] Coulee Dam, Thomas fled to Wellpinit with his father's bell.[7]

The story of the bell shows how the Spokane people were pushed and pushed from locations even when we tried to embrace white religion and schooling. Chief Garry had done his best to prepare his people for the white settlers. Nevertheless, in 1881 the reservation was created forty-eight miles northwest of the city of Spokane. Chief Garry was not pleased that almost all of the lands he grew up on, traveled, and lived on were not included in the reservation. Chief Garry refused to move to the reservation and died in Indian Canyon, on the outskirts of town. Finally, with the completion of the Grand Coulee Dam in 1946, the Bureau of Reclamation created Lake Roosevelt, and its slack waters flooded the land and the Spokane Indian River Church. The Grand Coulee Dam brought the traditional way of life of the Spaqni to an end. The lifeways of fishing, with a vibrant trading economy and a culture intertwined with the salmon, all but disappeared. Throughout her long life, my great-grandmother spoke of our lost cultural practices and shared the stories of our people, to remind us of our original tribal ways.

My tupiye Sadie was born in a teepee at the confluence of the Spokane and Columbia Rivers, an important historical site. Our people had always gathered there, but later on, the U.S. military would build their fort there to keep an ever-watchful eye on the Spokane and Colville Tribes. Despite colonial interference, our people remained deeply connected to the river. Not only did they begin their lives there, but they also ended their earthly journey there. We buried our people along the river and the riverine terraces, but this traditional practice ended with the rising waters of the Grand Coulee Dam.

As the construction of the dam began in the 1930s, officials relocated the white gravesites of small towns along the river. Tribal burial sites were not considered until much later. The Bureau of Reclamation finally commissioned Howard T. Ball to do all the exhumations, and he documented his work in the short film *Echoes of Yesterday* in 1939.[8] According to tribal preservation programs, the workers began with the white settler graves and only had a short time to finish the tribal burial exhumations. Ball hired 102 local tribal members to help with the excavation. In the film, Ball tells the story of how Colville chief Jim James, while driving his horse and buggy, came across some people pounding stakes into the ground. James asked them what they were doing. The white men replied, "Haven't you heard? We are going to build a dam and flood the area." A large part of the area to be flooded included reservation land. The tribal people objected, but the Grand Coulee Dam was built anyway. Trees were logged, some homes were burned or moved, and other homes were flooded.

According to Ball, this flooding required extensive exhumations of at least 1,388 burials of our ancestors. Workers subsequently placed these remains in four different cemeteries. The two new burial sites I am most familiar with are at Keller on the Colville Indian Reservation and at West End on the Spokane Indian Reservation. Along with Ball, John Bruce Dodd (1870–1945), one of the original founders of Ball & Dodd Funeral Home, also remarked in the film that moving the Indian graves to a new location on higher ground proved a difficult project. The relocation team removed graves from more than fifty different locations covering an area of 160 miles. Boats provided the only way to reach the remote gravesite areas along the river. The workers constructed wooden boxes at each camp area to hold the remains of our ancestors.

As I was watching the film, I heard the stereotypical Indian drumming playing in the background and felt irritated. The film's narrator, Gary Dixon, explained that as the crew worked and the graves were opened, the tribal people watched as the graves of their ancestors were desecrated. My heart sank when I heard these stark facts. Just imagine how they felt, having their ancestors dug up and removed from their intended resting place. Ball reports that many Indians even put up their teepees and watched the men as they worked. I suppose they felt the need to serve as somber witnesses, out of respect.

However, the makers of the film portrayed the exhumation and reburial as a happy time, with dancing and our traditional gambling "stick game" being played. I do not believe the filmmakers offered an accurate context of the intense sentiment regarding the forced reburial of our ancestors. Indeed, our tribal elders can confirm that the reburials started with the white settlements, with Native burials only an afterthought, which further shows the extent of white indifference. Ball himself had informed our elders that he would not have enough time to exhume all the tribal graves. Therefore, workers had to leave some of our ancestors buried along the riverbanks, and now their remains threaten to resurface with the regular rise and fall of Lake Roosevelt. Tribal cultural preservation crews must work to recover and rebury any newly disturbed remains.

During the excavation, Howard Ball and John Dodd asked tribal elders to pinpoint the location of the burials. Workers subsequently unearthed hundreds of artifacts, including an old flint lock brought to the area by the Hudson's Bay Company and subsequently adopted for use by the Native people. Excavators also found traditional spangles, earrings, arrow sharpeners, onyx chiselers, hide scrapers, copper spearheads, and arrowheads. Augusta Williams, a tribal elder who was 104 years old during the time of the excavation in 1939, told workers where to find the burials. Williams directed them to Kettle Falls, a major cultural, fishing, trading, and burial site for all the tribes in the area. Williams, in the film, modeled a recovered breastplate, which would have protected a warrior's body from arrows in battle.

According to the filmmakers, these artifacts supposedly became the property of Indian people. I find that conjecture highly unlikely. I have served on the Spokane Tribe's Cultural Advisory Committee and have no knowledge of any items returned to the tribe. In fact, many Indian artifact collectors and museums acquired funerary objects, despite their claims to the contrary. For example, Indianapolis resident Don Miller collected Native American artifacts illegally and improperly. In 2019 FBI agents found forty-two thousand items; half of them could be identified as Native American and constituted enough human remains to make five hundred people.[9] This practice also occurred here in Spokane when artifacts obtained from tribal families were handed from one museum to another. One such collection was located at the Schoenberg Center, established by Fr. Wilfred Schoenberg in 1968.[10] The building was designed to mimic a teepee and was created as the Museum of Native American Art and Culture. In the early 1990s, most artifacts inside the building were turned over to Spokane's Northwest Museum of Arts and Culture.[11] Very few of the artifacts have been returned to Native families, and not always with due diligence for accuracy. Our Native artifacts tell the story of how our people once lived along the river, but they continue to be misappropriated by non-Native amateur collectors and professional curators.

In the film, Dixon states that Indian ancestors were interred at the Kettle Falls location for four hundred to five hundred years. Workers discovered the distinct remains of a chief's baby, wrapped in a buffalo robe and adorned with drilled elk's teeth and copper beads. Excavators also found mass grave sites from what was believed to be a smallpox epidemic. Ball & Dodd's workers moved as many graves as they could before the water began to back up and the Columbia and Spokane Rivers flooded the bank shore. The team then had to abandon many remains to the rising waters.

Archaeologists who later learned of Howard Ball and John Dodd's reburial effort criticized the undertakers' methods and, in fact, called the entire operation a "major calamity" because of the seemingly indifferent destruction of scientific evidence. Apparently, Ball & Dodd workers ignored standard archaeological practices, such as careful handling and proper documentation, in service to the hurried nature of the project. Furthermore, according to the National Park Service, proper "archaeological investigations of the reservoir area were essentially an afterthought in the late 1930s," due to the rush to meet the timelines of construction at the Grand Coulee Dam and the corresponding need to clear out the Indian homes and towns located along the reservoir riverbank. When Howard Ball and John Dodd were contracted to relocate the graves, this sensitive endeavor became "strictly a burial relocation project, separate from the subsequent archaeological excavations."[12] During the exhumation of Indian burials, workers discovered hundreds of artifacts that had been buried with the deceased. Many of these cultural patrimonial and funerary objects were not reburied with the bones. Instead, they were collected for their intended return to the tribes, which we also now recognize as a problematic, if not downright deceptive, practice. Overall, Howard Ball and John Dodd unearthed our ancestors in an insensitive and unethical manner, and the destructive effects of the project still reverberate in our community today.

The building of the Grand Coulee Dam flooded the confluence of the Spokane and Columbia Rivers. As recounted by our tribal elders, it was as if ten thousand years of life and culture disappeared almost overnight. Ethnographer William Elmendorf noted that the Spokane people had often dug their graves in rockslides, and most of their "burials were found in sandy riverbanks one and one-half to four feet below the surface, with the heads of the deceased orientated downstream."[13] Others preferred sites within riverine terraces along the Spokane River, or often on the adjoining loose basalt talus slopes. While the Spokane Indians clearly took great care and respect with their burials in accordance with sacred tribal traditions, the creation of the Grand Coulee Dam ultimately forced the relocation of our ancestral remains in a hasty and disrespectful manner.

During the early 1900s, up until 1936, tribal members like Pierre Sockemtickem also grew crops and owned farms along the river. Our elders remembered the acres and acres of delicious cherries Sockemtickem had farmed. The building of the Grand Coulee Dam marked the beginning of wealth for some people in the Northwest, but it also foretold the beginning of the end for seventeen miles of river and everything the river had provided for the Spaqni. On these riverbanks, the Spaqni people had established homes, farms, and churches, and they had buried their ancestors there for many generations.

To this day, the Spokane Tribe has not been compensated for loss of land or fish. The river had provided a way of life and, thus, economic vitality for the Spokane Tribe. The building of the dam eradicated our cultural practices, economy of trading, and religious traditions. The Spaqni no longer had access to a free-flowing river or its salmon. Consequently, we also lost, for too long a time, the tribal ceremonies that accompanied the salmon migrations.

The Grand Coulee Dam constituted the greatest concrete structure ever built, until the Three Gorges Dam in 2009. But as the dam went up, so did the water. The project flooded the lands of everything upstream. Today, where what is now known as Lake Roosevelt, the water is drawn down and filled up to reengineer the river. Back then, white engineers called the dam a "man-made marvel" that had "the power to irrigate, create energy, and control flows."[14] However, this ongoing practice dramatically changes the landscape, as this constant lowering and flooding exposes the cultural resources of the Spaqni along the sacred locations of our ancestral remains. The waters along the river are flooded then released, which damages these sites as well as the land in general. In 2009 large chunks of Spokane tribal land (mostly sand embankment), nearly an acre in size, slid into the Spokane River, causing a wave to cross the river and slam into the recreation area at Porcupine Bay. Witnesses reported an eight-to-twelve-foot-high wave, a "minor tsunami," which overcame the boat launch at the Porcupine.[15] The reengineering of the river continues to cause landslides that threaten the stability of the riverbank. The tribe continues to bear the bulk of the negative burdens of the Grand Coulee Dam.

Overall, non-Indians downstream still benefit from irrigation and cheap electricity. However, the constant fluctuation of the water on Lake Roosevelt is a threat to environmental and cultural resources. The fluctuating lake level is hard on the ecosystem and destroys habitat. The Spokane Tribe's connection to the river and its banks cannot be overstated, and we, too, are devastated by the ongoing disruption via dam functions. My uncle Bruce Wynne, who served on the Spokane Tribal Council for sixteen years and as its chairman from 1989 to 2000, said, "We are river Indians. We lived on the river. What we had down there (on the river) and what was in place is desecrated every year. Every time there is a drawdown there is desecration. Graves come open. [Our traditional] camping sites come open and they are eroding. There is no payment for that. We have been willing to compromise. Then they ask us to give more. That's where the emotions come in. [We] are not going to give anymore."[16] Due to everyday dam operations, the Spokane Tribe must continue to suffer both the desecration and destruction of our Native lands.

The building of the Grand Coulee Dam not only permanently disturbed our ancestral burial grounds, but also changed the lived experience of the Spokane tribal people ever after. Other dams on the Spokane River, such as Upriver and Nine Mile, and all the way up to the one on Monroe Street, also deprived us of our traditional way of life. For more than one hundred years after the dams were built, we no longer fished for salmon or held our large, community-wide salmon ceremonies; nor did we canoe the river. Yet for thousands of years prior to the dams of the 1930s and 1940s, Indigenous people had traveled the rivers in canoe journeys that culminated in the salmon ceremonies. Since time immemorial, the Spokane people had gathered with other tribes to harvest the salmon and celebrate the return of the salmon at Kettle Falls. At this gathering, everyone did their part to help harvest the salmon that would be used for trading and to dry the salmon that would keep them fed during the long winters. Chief Garry was fishing for sustenance (and not just for sport) at Kettle Falls when a *suyapee* (or white man, which translates literally to "upside-down face") invaded his property and burned down his home near Hillyard, east of Spokane.[17] That incident symbolized the terrible changes to come for the Spokane Indians.

After the rising waters of the Grand Coulee Dam flooded our territories, local Indigenous people held the Ceremony of Tears in 1943, led by the Colville Tribe. As documented in *Echoes of Yesterday*, "The Indians came together one last time. To bid farewell to their fishing grounds and to reminisce. It was a sober [and] solemn occasion. The Indian people shared stories, experiences, and old legends about what had taken place on the shores of this mighty river for thousands of years."[18] The salmon would not be coming up the river again. This prediction proved true—we lost the annual salmon run. However, tribal elders and brothers, Martin, Albert, and Pierre Louie, continued to conduct the yearly salmon ceremony on a very small scale, albeit without the actual salmon, to pray for the waters of the Columbia and Spokane Rivers. When the Louie brothers passed, Martin's (now also late) son, Tom Louie, became the caretaker of the ceremony. Mike Finley, former Colville tribal councilman and tribal historian, also helped Tom keep the tradition alive. With mother-daughter duo Shelly Boyd and Stevey Seymour from the Inchelium Language and Culture Association, larger gatherings for the salmon ceremony began in July of 2016.

That year marked a revival of the traditional songs and prayers, as well as the feast and giveaway, which had not been held on this scale since the 1940s. Our ancestors had held the salmon ceremony to thank the Creator and give honor to the spirit of the salmon. For the Indigenous families of today, the salmon ceremony strengthens our connection to our ancestors and inspires our young people to embrace their culture and identity. Local tribal youth are once again returning to our rivers and traveling the reclaimed routes of our ancestors. They now join the canoe journey to camp for many days along the rivers and learn to navigate the waters. They start at different locations but congregate at the now-thriving salmon ceremony at Kettle Falls. Paddlers are greeted with Native songs, drumming, and war whoops.

For the first renewed canoe journey in 2016, the Upper Columbia United Tribes (UCUT)—which includes the Spokane Tribe, the Confederated Tribes of Colville Reservation, the Coeur d'Alene Tribe, the Kalispel Tribe, and the Kootenai Tribe of Idaho—secured five, four-hundred-to-eight-hundred-year-old cedar trees from the Quinault Indian Nation for each tribe to carve into canoes. This purchase of cedar trees renewed and updated the river canoe culture for these inland tribes. For the past five years, our people have also used the canoes to paddle throughout the Northwest to bring attention to water quality and to bring back the salmon. After the initial canoe carving, local tribes added one sturgeon canoe and three more dugout canoes for use by tribal paddlers. Tribes use the canoes for both traditional ceremony and environmental activism.

In 2018 we gathered for the canoe journey on the Pend Oreille River. Kalispel tribal elder Francis Cullooyah stated, "For the Colvilles, Spokanes, Kalispels, Coeur d'Alenes, and Flatheads, it has always been a dream of getting back out on the water. On our rivers." Cullooyah spoke in depth about how the canoe journey can empower the tribes:

> In our way of thinking, in our way of life . . . this Canoe Journey is like our winter dances because it brings all the tribes together in one place. To me it's a blessing and it gives me good feelings to be a part of this journey. Little did I know that one day I would be standing here talking to you and speaking in this manner. This is important for you as a young person. Many of you young people that come here—you are building your strength! You are building your life! One day that strength will be there in your prayers and your thoughts. One day, when life comes, and things get tough—that strength will be with you. Our Creator will be watching over us.[19]

Reclaiming our traditional waterways is making our young people strong and reminding them of our connection to the rivers. My children are participating in these canoe paddles and taking part in these ceremonies. These traditional practices reunite our people across generations and tribes.

The tribes begin their canoe journeys in various locations but always end them at Kettle Falls for the salmon ceremony. For example, the Inchelium canoe called Snklipqn launched in 2016 from Lower Arrow Lakes, Syringa, British Columbia, Canada. The Canadian government had declared the Sn̓ʕay̓ckstx (Arrow Lakes) Band "extinct" in 1956, only for that decision to be reversed in 2021 after a ten-year court battle.[20] Prior to this legal victory, even though the Canadian government did not recognize the Sn̓ʕay̓ckstx Band, they still maintained their sacred spiritual practices by participating in the paddle.

Each year the salmon ceremony begins with an explanation in Salish dialects accompanied by a translation into English. Leaders from each of the tribes acknowledge all the hard work of not only those who first carved the canoes from the cedar trees, but those who now take up the oars to paddle as well. They also thank all of those who support the paddlers with food and camp preparations. At the gathering, we celebrate our traditions and our languages. We focus on bringing the salmon back to the rivers and healing our people and our waters. Our people once thrived on the river's fish in conjunction with the land's roots and berries—our first foods. Today, we strive to bring back the salmon as both a nutritional *and* spiritual imperative.

On July 22, 2018, Richard Armstrong of the Caylx Tribe spoke from the riverbank to the assembled tribes: "We gather for the salmon ceremony, for our people and for the paddlers." He asked us to pick up rocks and pound them together: "And [with the rhythm of these rocks] we speak the language of the salmon. What we are doing is from our hearts. We are working to maintain the sacredness of our waters. As tribal people we carry on what our ancestors did. To do this work—we keep in our memories the ones who took care of the rivers—we continue the ways of our ancestors. We sing the Sturgeon Song. We sing the Salmon Song."[21] Armstrong also said that we should remember the words of the late Martin Louie: "The salmon know no boundaries. They travel the world to all our people. When we pray—we pray for the sacredness of the river, and we pray for our river to be clean. We pray for the animals. For the children that are not born yet." As Indian people we know it is our responsibility to pray and sing for the river. As Indigenous people we need to do these ceremonies for the river and be there to support it. The river has always been there and supported us.

The Spokane Tribe has fought to protect both the cultural resources and water quality of the Spokane River. As the attorney for the Spokane Tribe for more than ten years, I collaborated with Spokane Tribe's Natural Resources director Mary Verner and tribal attorney Dave Lundgren, who both worked for the tribe to draft the application for Treatment as a State

(TAS) under the Clean Water Act. The tribe was approved for TAS, and our Spokane Tribal Natural Resource Department set some of the highest water quality standards in the state. These new water quality standards aim to provide clean, cold, in-stream flows that will enable fish to survive.

The dams tragically changed our way of life, and we almost lost forever the traditional cultural practices and ceremonies our ancestors had observed for thousands of years. Today, our tribal biologists and hydrologists collaborate with non-Native allies to purchase land and reintroduce salmon in areas that the fish have not been able to reach since the 1930s due to the dams. The tribes currently work with the Washington Department of Fish and Wildlife (WDFW) and the Inland Northwest Land Conservancy (INLC) to prioritize restoration on the Spokane and Little Spokane Rivers. Most recently, on August 6, 2021, the local Indigenous people, collectively known here as the Sqelixw, came together at the Little Spokane River along with our partners at the WDFW and INLC. From the riverbanks, our people spoke of how our grandparents suffered the loss of their lands, homes, and staple food source, the sacred salmon, in addition to their ancestral burial grounds, with the building of the dams.

Next, we released fifty-one summer Chinook into the Little Spokane River, an effort made possible by INLC's $1.5 million purchase of "95 acres—including 1,700 feet of shoreline—west of the Little Spokane River in 2020," right next to 104 acres of WDFW land.[22] For this initial release (more are planned), tribal biologists tagged the salmon with radio transmitters that can track the fish in real time. This information will help with the protection and restoration of urban rivers, like the Spokane, to promote salmon reintroduction. Tribal biologists also hope the salmon will spawn and build nests in which female salmon can lay their eggs. The UCUT have been diligently working to restore water quality and reintroduce salmon into selected areas. The radio trackers will monitor how the fish make their way downstream and illustrate exactly how the fish can survive going through or over each dam. The tribes and their partners anticipate a permanent passage from the Spokane to the Columbia, all the way to the Chief Joseph Dam.

Traditionally, the salmon chief always selected a young Spokane Indian man to sing a song to call the salmon home. On August 6, 2021, for the first time in 111 years, a Spokane-Colville Indian man, my cousin Isaac Tonasket, sang a song for the salmon. Tim Peone, tribal fish biologist, noted that the salmon splashed wildly in their holding tank as Tonasket drummed and sang the chief's song. On this momentous day, our people gathered in a circle with non-Indian scientists, the WDFW, INLC, the Spokane Riverkeeper, and many others as we danced our traditional "Spokane Indian Happy Dance." We hope to hold additional events to educate others and honor the culture of our ancestors. We look forward to a future of sustainable reintroduction of salmon into the Upper Columbia and Spokane Rivers. We will persist in our coordinated efforts to protect the watershed and habitat. As our tribal elders remind our young people, returning to the river and practicing the ways of our ancestors will make us strong. Stevey Seymour of the Sn̓ʕay̓ckstx (Arrow Lakes) Band reminds us, "It is our spiritual responsibility to [work for] the healthy return of the wild salmon on the upper reaches of the Columbia [and Spokane Rivers]."[23] We must remember the sacrifices of our ancestors, as well as the sacred songs and ceremonies that carried them and the salmon of the Sƛ̓x̣etkʷ, our "Place of Fast-Moving Waters."

Federal agencies reengineered the Spokane and Columbia Rivers to benefit the region, but our people paid the biggest price for such "progress." We lost our lifeways: homes, farms, churches, and ancestral burial grounds, in addition to the life-sustaining river and its salmon. Ongoing human manipulation of the river waters today negatively impacts the ecosystem and sacred lands of the Sqelixw. We buried our people along the river, and though workers reburied many on higher ground during the building of the dams, our sacred lands continue to be desecrated as our ancestral remains resurface. Yet, through the dedicated and diligent revival of our culture, stories, and ceremonies, our people demonstrate resilience and maintain our connection to the rivers. We are once again reclaiming our waterways with our canoe journeys and salmon ceremonies, as well as by advocating for clean water and the return of the salmon to local waters. We are keeping our sacred promise to our rivers, and we will continue to sing the songs until the salmon return home.

NOTES

1. Clark, *Indian Legends*, 74–75.
2. Charles Byers, "Carving the Columbia: The Ice Age Floods," Spokane Historical, Eastern Washington University, accessed August 18, 2021, https://spokanehistorical.org/items/show/656.

3. Margo Hill and Warren Seyler, "Spokane Tribal History, Federal Indian Policy & Law" (lecture at Washington State University Health Science Campus Leadership, Spokane WA, September 29, 2021).
4. "Sadie Boyd Dedicates Monument to Chief Lot," *Rawhide Press*, November 1969.
5. Dave Archambault Sr., in discussion with the author, July 19, 2021.
6. "Historic Garry Bell Part of Indian Memorial Rite," *Spokane Daily Chronicle*, July 20, 1973.
7. My great-grandmother Sadie Boyd had this story inscribed on the monument located at the Wellpinit Presbyterian Church Cemetery.
8. Lon Gibby, "Echoes of Yesterday," 1939, accessed October 2, 2021, YouTube video, 20:31, https://youtu.be/92-SGupFRqU.
9. Domenica Bongiovanni, "What Happened to the Thousands of Artifacts, Human Bones the FBI Found on Indiana Farm," *IndyStar*, February 27, 2019, https://www.indystar.com/story/entertainment/arts/2019/02/27/what-happened-don-miller-artifacts-fbi-found-indiana-farm/2990542002/.
10. Jesse Tinsley, "Then and Now: Museum of Native American Cultures," *Spokesman-Review*, November 2, 2020, https://www.spokesman.com/stories/2020/nov/02/then-and-now-museum-of-native-american-cultures/.
11. "Indian Museum to Shut Doors; Collection Will Stay in Spokane," *Seattle Times*, November 17, 1991.
12. McKay and Renk, *Currents and Undercurrents*, 54.
13. Ross, *The Spokan Indians*, 181.
14. Bureau of Reclamation, "Grand Coulee Dam: A Man-Made Marvel," May 13, 2014, YouTube video, 42:36, https://youtu.be/BU4qw9zYX9Y.
15. "'Minor Tsunami' Closes Lake Roosevelt Campground," *Spokesman-Review*, August 26, 2009.
16. Bob Bostwick and Warren Seyler, "Spokane Tribe—A River of Promises," 2000, YouTube video, 28:33, https://youtu.be/jHztZQ7Ak-w.
17. Shelly Boyd, in discussion with the author, August 12, 2021; Jessett, *Chief Spokan Garry*, 202.
18. Gibby, "Echoes of Yesterday."
19. Margo Hill, "Tribal Canoe Journey—'We are One'—Kettle Falls Salmon Ceremony," *Rawhide Press*, July 2018.
20. *Her Majesty the Queen v. Richard Lee Desautel* [2021], SCC 38734, accessed October 14, 2021, https://www.scc-csc.ca/WebDocuments-DocumentsWeb/38734/FM010_Appellant_Her-Majesty-the-Queen.pdf.
21. Hill, "Tribal Canoe Journey."
22. Eli Francovich, "It's Been over a Century, but Summer Chinook Are Back in the Little Spokane River: 'It's Kind of a Spiritual Experience,'" *Spokesman-Review*,

updated August 11, 2021, https://www.spokesman.com/stories/2021/aug/06/its-been-over-a-century-but-summer-chinook-are-bac/.

23. Derrick LaMere, "Calling the Salmon Home," 2021, accessed October 13, 2021, YouTube video, 15:37, https://youtu.be/8mmm43FFpLo.

BIBLIOGRAPHY

Clark, Ella. *Indian Legends of the Pacific Northwest.* Berkeley: University of California Press, 1953.

Jessett, Thomas E. *Chief Spokan Garry, 1811–1892: Christian, Statesman, Friend of the White Man.* Minneapolis: T. S. Denison & Company, 1960.

McKay, Kathryn L., and Nancy F. Renk. *Currents and Undercurrents: An Administrative History of Lake Roosevelt National Recreation Area.* Washington DC: National Park Service U.S. Department of the Interior, 2002.

Ross, John Alan. *The Spokan Indians.* Spokane: Michael J. Ross, 2011.

3 Terms and Conditions Apply

Greg Gordon

To celebrate Earth Day 2017, about thirty-five people hunched over a twenty-by-twenty square-foot mat of recycled plastic and burlap, inserting various wetland plants into predrilled holes. After packing the plants with potting soil, they lugged the buoyant mass to the water's edge where a rowboat waited to haul the artificial island to the middle of a small lake and anchor it with a steel cable attached to a concrete block. Planted with cattails, sedges, spirea, and iris, this tiny island would provide habitat for painted turtles, waterfowl, and other species. But perhaps more important, it would absorb excess nutrients, such as phosphorus and nitrogen, to help alleviate the algae blooms that occur each summer in this former slough of the Spokane River.

This floating wetland was the culmination of three years of effort by students and faculty at Gonzaga University, a midsize liberal arts university, in Spokane, Washington. The campus lies just a few blocks from downtown and is bordered by busy streets, railroads, repurposed warehouses, and grain elevators—remnants of an industrial past. Brick buildings, parking lots, and lawns are the primary landscape features, and these largely define and constrain student activities. Deliberate landscaping sculpts decorative flower beds while towering trees provide shade for students in plastic lounge chairs. Like most campuses, there are a few "natural" features, neglected or on the margins of the campus: a wetland, stream, or meadow, often tucked behind the science building.

In this case, Gonzaga borders the Spokane River on its journey through the city. Between the main campus and the river lies a pond, euphemistically called "Lake Arthur." The campus lawns slope away from the buildings toward the lakeshore, which is primarily dominated by Canada geese and students in hammocks. As a human artifact, it can hardly be considered wild, even natural is stretching the term. Yet for many students, Lake Arthur is their most intimate encounter with the natural world during their college career, and they are actively attempting to improve its ecological function and value. Does this artificial pond then signify a new and emerging view of nature in the twenty-first century, one in which the human and natural worlds merge, coexist, and even enhance each other?

Over its forty-year history, Lake Arthur has shifted ecological states, which, not coincidentally, correspond to Americans' changing perceptions about the natural world. For much of the nineteenth century, America's dominant culture contained two conflicting views of nature: those of Manifest Destiny and the Romantic movement.[1] These two views collided in Spokane during the 1880s, when the Jesuits purchased 320 acres from the Northern Pacific Railway to build a college along the north bank of the Spokane River on the ancestral homeland of the Spokane Indians. The Jesuits focused on the aesthetics of the site, shaped as they were by a spiritual tradition that recognized the inspirational power of beauty and nature. Their purchase included a bay, or large back eddy, formed by a peninsula jutting into the river, and the fathers regarded the location as conducive for contemplation and study.

But in 1887, just weeks before the college was to open, "suddenly a railroad appeared before the fathers like an ugly blight" as the Oregon Railroad and Navigation Company began constructing "a hated roadbed between the college and its lovely bay . . . not the last, nor most disturbing to the peace of the Jesuits."[2] Two years later, the Jesuits took to the courts to stop a second railroad along the north side of the bay but lost, as the right of way fell under eminent domain. Unable to overcome the forces of industrialization, the Jesuits abandoned the serenity of the bay and moved the buildings farther inland and away from the railroads and the river.[3]

The physical shift in campus orientation mirrored a shift in cultural values from Romantic to utilitarian views of nature in the early twentieth century. The Jesuits now had little use for the lower campus, and in

1904 leased the land to a lumber mill. Two years later, five railroad lines bisected the campus with several trains per day dumping saw logs into the bay, which had become a mill pond, partially separated from the river by the railroad bed. By 1922 five hydroelectric dams along the river divided the river into a series of slow-moving reservoirs, and completion of the Grand Coulee Dam in 1942 eliminated salmon runs from the entire upper Columbia River watershed, including the Spokane River. The loss of salmon, the river's keystone species, had a cascading effect on both Native peoples and the ecosystem.

In reaction to the excesses of industrialization and utilitarianism, a rising ecological consciousness permeated American culture in the 1960s and '70s. Spokane and Gonzaga were no exception and embraced the nationwide beautification effort. Spokane began preparation for the first environmentally themed World's Fair—Expo '74—by dramatically restructuring its downtown river corridor from an industrial railroad yard of warehouses "into a scenic asset" around the theme, "Celebrating Tomorrow's Fresh New Environment."

Concurrently, Gonzaga's vice president Fr. Arthur Dussault led a decade-long campaign to beautify the lower campus, which had become a dumping ground and eyesore after a fire destroyed the lumber mill in 1945. In the early 1960s, students began a series of litter cleanups, and Dussault raised the funds to convert the former mill pond into a campus park complete with a fountain, intensive landscaping, and a causeway that separated the pond from the river.[4] Although Lake Arthur and the surrounding manicured lawn improved the aesthetics of the lower campus, it attracted little more than a passing flock of Canada geese, which contributed to the lake's nutrient load and subsequent algae blooms. But by the twenty-first century, painted turtles began showing up in the pond, along with a pair of beavers, occasional great blue herons, and wood ducks. With the accidental recolonization of native species, the state of Washington identified Lake Arthur as the only wetland on the Spokane River within the city limits.

As a human-created ecosystem, Lake Arthur is hardly unique. In an era in which humans have impacted every ecosystem on Earth—in other words, the Anthropocene—ecologists recognize four ecosystem states that often occur along a continuum. Serving as the baseline, natural, or *historical*, systems function free of human manipulation. But when humans push those ecosystems beyond the range of historical variability—by say logging forests, grazing livestock, allowing activities that result in increased flooding or fires, or introducing invasive species—a *hybrid ecosystem* results. Such hybrid systems retain characteristics of natural systems, but the species composition and ecosystem functions have shifted outside their historic range. Yet, restoration is still possible. But once a system crosses an ecological threshold as a result of human action, such as a drastic change in hydrology or the elimination of keystone species, so that a return to the original state is unlikely, it becomes a *novel ecosystem*. Dams and channelization, for example, significantly alter a river's hydrology, sedimentation, and nutrient load, while the placid waters of the reservoirs favor non-native fish. This combination of changes in function, structure, and species composition cause a river to pass beyond the range of historical variability into a new ecological state—a novel ecosystem. In some cases, such as with the Elwha River in western Washington State, the singular act of dam removal can reverse the process leading to a restored ecosystem, but this is nearly impossible with urban rivers that have been so heavily impacted by a multitude of factors. At the opposite end of the spectrum, *designer ecosystems*, such as a city park, golf course, or wheat field, are a result of intentional human manipulation of a system's structure and function and require continual maintenance.[5] Interestingly, Lake Arthur has experienced all four ecosystem states, each corresponding with a particular time period and reflecting society's prevailing view of the natural world.

Not surprisingly, novel ecosystems often occur in "urban regions with heavily altered abiotic conditions and large numbers of exotic species."[6] Indeed, geographers now consider approximately one-third of the planet's ice-free landmass to consist of such novel ecosystems. Invariably dammed, channelized, dewatered, filled with exotic species, and beset by excess nutrients from agriculture and industrial pollution, urban rivers are quintessential novel ecosystems. While urbanization threatens more species in the United States than any other human activity, primarily through habitat loss and replacement of native species with non-native, novel ecosystems can provide ecological and human benefits. These new communities of native and exotic species functioning in a dramatically altered environment can potentially become self-sufficient and self-sustaining ecosystems, wild even.

Historically, urban rivers are the most neglected parts of a city. In straightjacketing the Spokane River and its tributary, Hangman Creek, the city dumped broken blocks of concrete, or "urbanite," along the banks for erosion control. At first, this detritus of industrial civilization appears unsightly, but a closer look reveals that urbanite forms the infrastructure for an accidental ecosystem, in much the same way as a sunken ship can provide the foundation of a coral reef. Urbanite houses marmots, which, in turn, bring coyotes, badgers, red tail hawks, golden eagles, and even an occasional cougar. Urbanite also provides housing for ground squirrels, porcupines, and raccoons. Knapweed, knotweed, hawthorn, Russian olive, ponderosa pine, and cottonwoods all grow out of the jumbled concrete blocks. With a mix of old and new species, urbanite provides the foundation for a new ecosystem in the middle of the city—a stark contrast to the manicured and fertilized grass of city parks. While on one hand, such hybrid and novel ecosystems present many challenges, they might also offer our best hope for the future of biodiversity in the Anthropocene.

Addressing urban ecology becomes increasingly important as urbanization gobbles up habitat. As science journalist Richard Conniff notes, the rapid decline of bird species has "made it disturbingly evident that it's not enough for cities to plant a million trees, preach the gospel of backyard gardens, or build green roofs and smart streets." Cities need to provide some sort of "green infrastructure" as space for "breeding, shelter, and food."[7] Ideally, this habitat would be along wildlife corridors or "stepping stones"—patches of habitat easily connected by wildlife moving through a matrix of asphalt and lawns. Lake Arthur is a prime example of such a stepping stone. Urban ecosystems often appear as a mosaic "in varying states of modification."[8] The connectivity of such patches is vital to a landscape approach when seeking functional connectivity. In this manner, such patches can also serve as "urban arks," maintaining species, albeit in artificially dense populations.

In urban environments, novel ecosystems can play an important role in biodiversity conservation and landscape connectivity, both as habitat refugia and as corridors.[9] One way to make effective use of our conservation lands, which are increasingly fragmented, is to provide corridors so species can move from one core habitat to the next. Climate disruption makes this even more crucial, as both plants and animals will need to migrate to more suitable habitats, either temporarily or permanently. Rivers and streams—channeled, ignored, or relegated to industrial disposal systems for much of the twentieth century—contain both the highest biodiversity and potential for rewilding.

Unintentionally, the creeks and rivers flowing through our cities function as de facto wildlife corridors, bringing plants and animals into the city and providing escape routes as well. In other words, urban rivers can help mitigate the biodiversity crisis. For example, much of the riparian diversity along the Spokane River appears accidental, as herons, kingfishers, wood ducks, and otters hug the shorelines, probing, darting, and paddling among the willows and blocks of concrete riprap. As novel ecosystems, we can allow these systems to reshape themselves, and if we do it right, we allow species to experiment with new adaptations and associations and form new ecosystems.

Interestingly, there seems to be a peak in species richness close to the midpoint of the continuum between urban and rural. On the suburban fringe, for example, intermediate disturbances create "spatial heterogeneity," and human activity can boost the availability of scarce resources, such as water, nutrients, and habitat.[10] According to geographer Michael McKinney, the increased species richness in these fringe areas results from the convergence of native with non-native species, especially cultivated plants, creating a hybrid ecosystem. Urban avoiders, such as large predators, admittedly might be missing from such an ecological assemblage. On the other hand, urban adapters, like coyotes, robins, and lizards, along with urban exploiters—mice, pigeons, and peregrine falcons—potentially could lead to a slight increase in species richness. While rare and vulnerable species often suffer with urbanization, in some cases novel ecosystems can provide crucial habitat for such species.

Traditionally, ecological restoration focused primarily on "ecological integrity and historical fidelity," or "intentionally altering a site to" its historical, precolonial condition and eradicating newly arrived non-native species.[11] A focus on the Anthropocene, however, allows us to examine the long history of human manipulation of nature. Whether it be the pre-Columbian earth works of Cahokia, or the extensive crossbreeding that turned a grass (teosinte) into corn allowing humans to plant it from Bolivia to New England, or the widespread use of fire by Native peoples to modify ecosystems to their benefit, suggests that few ecosystems on earth haven't been modified by people for thousands of years. In addition to the historical recognition of humans shaping ecosystems, the Anthropocene has also forced restoration ecologists to concede that climate change is fundamentally restructuring ecosystems, and the historical baseline might end up as an historical artifact rather than a functioning ecosystem.

Over the last twenty years, the recognition of the historical, present, and future role of humans in shaping ecosystems led the Society for Ecological Restoration to abandon its commitment to return ecosystems to historical conditions and instead advance "ecological restoration to sustain biodiversity, improve resilience in a changing climate, and re-establish an ecologically healthy relationship between nature and culture."[12]

We have many terms for what is happening to our urban riverscapes:

Restoration implies a return to a former state, as in restoration of a wetland by removing an invasive species, or restoration of a salmon stream by taking out a dam. Restoration is not usually an option for urban rivers, which are often impounded and channelized. Any functional restoration would be constrained by the physical infrastructure of the city itself. At the opposite end of the spectrum, *renaturalizing*, a somewhat awkward and vague term, usually refers to creating some natural amenities such as green space and native plant gardens. Renaturalizing describes a pseudo nature, a designer ecosystem that requires continual human maintenance. *Daylighting*, the process of ripping up pavement or some other urban infrastructure to expose a long-hidden creek, often occurs in concert with renaturalizing.

Revitalizing is probably the most apt description of the current effort surrounding urban rivers. We are essentially restoring the ecological and natural vitality to these ecosystems, and while this is an appropriate description and encompassing term, it lacks the imaginative weight of rewilding.

Rewilding not only captures the imagination but suggests a promise. It also recognizes that urban rivers are usually novel ecosystems and self-functioning. The premise of this book explores how we might allow nature itself to dictate the terms of engagement, with some ecological or structural nudges here and there. Such an approach often requires interventions but not maintenance. Rewilding can be a blending of ecological and philosophical approaches, the recognition that there are limitations to ecological restoration in urban environments, but that we can mix the pragmatic with the idealistic efforts to design and enhance systems so they can ultimately be self-sustaining and self-regulating.

The accidental rewilding along the Spokane River stands in contrast to deliberate renaturalizing, which is characterized by heavy investments in infrastructure to create green space out of urban blight. Examples of high-profile riparian renaturalization efforts can be found in two of the world's largest cities—Seoul, South Korea, and Los Angeles, California. In Seoul the $380 million Cheonggyecheon Restoration Project required the demolition of a highway to reveal a 3.6-mile stretch of creek through a densely populated and highly polluted section of the city. The project took five years and resulted in a seven-fold increase in overall biodiversity. Plant species jumped from 62 to 308, fish species from 4 to 25, insect species from 15 to 192, and mammals and amphibians doubled. The project also reduced the heat-island effect by six to eight degrees Fahrenheit, reduced air pollution by 35 percent, and attracts sixty-four thousand visitors to the creek every day—a social and ecological success by any measure.[13]

In the United States, the Los Angeles River is perhaps the best-known example of renaturalizing. Envisioned nearly thirty years ago, restoration of the Los Angeles River from a concrete viaduct into some semblance of a free-flowing river has gained significant traction in the last few years. The 2014 agreement between Los Angeles and the Army Corps of Engineers developed a $1 billion restoration plan for an eleven-mile stretch of river. The plan, yet to be implemented, would replace the concrete retaining walls with terraces, wetlands, and parks. The plan also includes bike paths, pedestrian bridges, public art installations, and riverfront restaurants.[14] (In the last chapter of this book, Bryan Rasmussen illustrates how a rewilding alternative to gentrification is unfolding on the Los Angeles River.)

As green space, such as walkways and bike paths, makes neighborhoods more attractive, it can lead to higher rents and housing prices, forcing lower-income households to move to less desirable areas, increasing the alienation from nature and simultaneously encouraging the commodification of nature. Such gentrification is especially evident along urban waterfronts as former industrial zones are remade into networks of retail shopping, restaurants, parks, and promenades. (C. Ian Stevenson's chapter on the Saw Mill River addresses this very dilemma.) Replacing the highway in Seoul with a walkable greenway caused nearby homes to sell at a premium after years of going for bargain prices. We are witnessing a similar process in Spokane with the redevelopment of a brownfield site into the city's hippest new neighborhood overlooking the Spokane River. Undoubtedly, manicured nature is an economic asset.

Certainly, gardens along streets and bike paths are positive developments, but green space is not necessarily habitat. These renaturalization projects require continual human maintenance and a high degree of species selection. With all the landscape planning and design, management and manicuring, isn't some of the essential "wildness" lost, the "self-willed land" that is just as imperiled by urbanization as individual species? What about the spaces between places, the so-called vacant lots, the feral areas between strips of land and forgotten creeks? What would happen if we allowed plants and animals to mix and re-form new assemblages, possibly creating wild, self-sustaining, new ecosystems?

Cities, of course, are not built with native ecosystems in mind. For much of the nineteenth and twentieth centuries, as humans began to shift from rural existence to urban life and cities exploded in growth, Americans viewed nature and cities as inherently contradictory. Cities, while full of economic opportunity, were evil, unsanitary, full of vice and mayhem. Nature, on the other hand, was pure, unsullied, and moral. With the rise of new urbanism, we seem to be making our peace with cities and transforming them into places that nourish our social and mental well-being. While urban planners increasingly incorporate parks and greenways into a city's infrastructure, they still suffer from a traditional conception of landscaping as even native plant gardens are cultivated and manicured to appeal to a pastoral aesthetic. Laid out in a grid and built along orderly straight lines by rational planners, cities are the antithesis of the natural world, and most municipalities regard wildlife as a nuisance. Whether it be geese, deer, skunks, or even bears, we employ a variety of means to keep wildlife at bay in our cities. And yet the critters keep creeping back—an accidental rewilding.

In the midsize city of Spokane, bald eagles, ospreys, and red-tailed hawks ride the thermals rising off the city infrastructure. Cooper's hawks zip through the pines searching for the insouciant quail cackling from the tops of garages. Wild turkeys strut through backyards and block traffic on side streets, so much so that residents' complaints have led to a search-and-destroy egg hunt to control the population. City-dwelling racoons and porcupines raise the ire of pet owners. Deer, and even moose, invade flower gardens. On "a fairly normal year," there are seventy-five "significant" moose incidents in the Spokane area, including sixteen relocated and four euthanized. And every so often a black bear shows up for a few weeks before deciding that the dogs and people are too much.

One evening while stopped at a traffic light on the city's main drag, I watched a large, healthy coyote trot across four lanes of intermittent traffic. He angled through the parking lot between Dutch Brothers Coffee and Zip's Drive-in before disappearing in a maze of alleys. Another time, I spotted a fawn in an alley one morning. Not more than a few days old, it was curled up next to a trash can, right where its mother had left it, while she browsed on shrubbery and newly planted gardens. Applying ancient evolutionary strategies to new environments is the mark of a successful species, and white-tailed does tend to return to sites where they have had previous success in raising young and may have even traveled miles to this secure but seemingly "unnatural" location. How are plants and animals (including humans) adapting to new habitats and creating novel ecosystems in the process? Are we witnessing the evolution of a new suite of species in addition to new behaviors? To what extent are climate change and land-use disruptions reassembling ecosystems, and might these new assemblages be more resilient than the ones they have replaced? If so, what does that mean for conservation in the twenty-first century? And are these new ecosystems any less vital or beautiful because they are an unintentional product of humankind? Are we creating a functional hybrid or a Frankenstein's monster?

Much of our current ecological crisis stems from how we think the natural world ought to look—we humans are the ones who find an urban coyote or deer incongruous. We need to shift our focus and try to see the landscape through the lens of the animals themselves. Like ourselves, animals and plants seek places (habitats) that provide resources and diminish threats. Ideally, urban rewilding can increase the former and decrease the latter.

Until recently, rewilding has largely focused on restoration of pastoral or abandoned landscapes and not so much on cities. Rewilding advocates assume a landscape level approach, such as in restoring wolves to Yellowstone or bringing back ancestral wild cattle to the Netherlands, not some impounded industrial sewer race through a warehouse district and filled with toxic contaminates. Admittedly, it might come across as a bit of a stretch to apply the term rewilding to urban rivers, but if we think about what really needs to happen to address our ecological and cultural crises, rewilding is more apt than any of the other words we use to frame this twenty-first-century transformation. Indeed, a small but growing movement is spreading across North America and Europe. For example, the Wild Cities Project is an international network designed to establish guidelines for nature protection in urban areas. Its goals are to protect and enhance ecosystem services as well as open space and recreational opportunities with a vison of keeping half of the ecosystem intact through protected parks, forests, and waters.

Many cities in Europe address urban wildlife from an ecological viewpoint, integrating wildlife corridors and habitat for native species in urban planning. While the United States has a system of urban wildlife refuges and programs to encourage migratory bird conservation, and many cities have urban habitat acquisition and management programs, nearly all suffer from lack of interest. Indicative of our priorities, only tiny fractions of university wildlife budgets are devoted to urban wildlife studies. This imbalance seems to be shifting, however, with an increased interest in urban ecology among biologists. Tragically, however, this might have more to do with the loss of wilderness study sites than anything else.

Nevertheless, there is a growing emphasis on urban wildlife and forestry. The U.S. Forest Service hosts a research center for urban forestry and recently developed iTree software to help foresters better manage urban forests for wildlife considerations. This change coincides with the emergence of academic journals dedicated to urban ecology. The lay person can now find urban wildlife guidebooks and online forums on urban biodiversity. For example, the Biophilic Cities Project (www.biophiliccities.org) seeks to connect urban dwellers to urban nature.

But overall, urban rewilding is still largely accidental with plants and animals using the empty spaces, the niches, that we've neglected. As paleoecologist Stephen Jackson notes of Gary, Indiana, "Trees grow in the ruins of factories, marshes reclaim railroad beds, and birds roost in collapsing houses. We are witnessing what might be called an Anthropocene rewilding, as industrial, urban, and agricultural lands are abandoned and novel ecosystems emerge."[15] Such accidental rewilding might include non-native species, challenging our notions of what we think the species composition of an ecosystem should be.

Rewilding places a high value on resilience, which in many cases might be a recombination of exotic species, generalists, and native specialists. Some ecologists have even begun to question the orthodoxy of prioritizing native species in areas of dramatic changes to the local ecosystems. Although it's still causing apoplexy among environmental purists, ecologists Donald Dearborn and Salit Kark acknowledge an increasing role for exotic species. They note that in terms of ecosystem services, exotic species "may be equivalent to, or sometimes even better than native species" in providing carbon sequestration, pollination, and pollution reduction. Honeybees and earthworms are examples of exotic species that perform vital ecosystem services. In Puerto Rico, ecologist Joseph Mascaro found that novel ecosystems often match or exceed native systems in nutrient cycling and biomass production.[16]

Perhaps, as Mark Davis, author of *Invasion Biology* notes, instead of worrying about the origins of species, we should identify those causing problems according to our values and deal with them. Native species can often function as aggressive invasives in certain times and places, while exotic species can function as "proxies" to recreate ecosystem processes once fulfilled by now rare or extinct species.[17] Who knows? Maybe Russian olive (often considered an invasive species) will provide native pollinators with their primary source of nectar one hundred years from now?

Living in the Anthropocene requires new ways of thinking about and interacting with the natural world as the human population and its ecological footprint continues to increase. Despite the 2022 United Nations agreement to conserve 30 percent of the planet for biodiversity by 2030, only 15 percent of the earth's surface currently lies in wilderness and national parks. As the total size of urban areas now exceeds that of such protected areas, we need to rethink cities and ecosystems so that we can create shared space that allows nature to evolve. The first step in rewilding is to view cities as habitats, however fragmented, rather than solely human endeavors of concrete and steel. In an urban environment, we have multifunctional landscapes composed of hybrid and novel ecosystems, which occupy the zone between natural and intensively managed ecosystems and can meet a variety of social and ecological goals. But to retain essential wildness, we need to recognize and restrain the human instinct for meddling.

While rethinking conservation in terms of urbanite—an unintentional byproduct of urban redevelopment and beautification—suggests benign neglect, it's more of a reorientation of our priorities that is needed. Along Hangman Creek, such realignment is beginning to take shape as the neighborhood council and city, with assistance from local landscape architects and the National Park Service, are developing a revitalization plan that would balance ecological enhancement, recreation, aesthetic appreciation, and cultural and environmental awareness of place. Rather than further disturbing and modifying the floodplain, the plan calls for integrating the existing industrial infrastructure. Trails would follow the gas pipeline and the tops of levees. Footbridges to cross the creek would use old concrete pilings and hang below the railroad bridge.[18]

Much of Hangman Creek has been channeled and lined with riprap—remnants of old buildings dumped along the bank to confine the creek to its bed. Rather than remove this urbanite, now necessary to prevent erosion, the revitalization plan calls for using it as anchors for native shrubs and other vegetation. With input from the Spokane and Coeur d'Alene Tribes, cultural resources, like edible and medical plants, will be interspersed amongst the urbanite, so that people can gain a tactile and gastronomic experience of place. Thus, the landscape becomes more than scenery, but also a source of sustenance—physical, cultural, and spiritual. For example, in 2021 the Coeur d'Alene Tribe, whose reservation lies at the headwaters of Hangman Creek, purchased forty-eight acres of land slated for development along the lower reach of the creek. The tribe plans to preserve and restore the property, with the ultimate goal of salmon reintroduction, demonstrating that cultural and ecological restoration can occur simultaneously.[19]

Urban rewilding potentially addresses two interrelated problems: biodiversity conservation and environmental justice. Less than a quarter of Americans live in a rural environment where nature is something tangible, while more than half of the human population currently lives in urban centers, projected to increase to 68 percent by 2050.[20] In the continental United States, 80 percent of us live in cities or suburbs, and most of the earth's inhabitants can scarcely afford to visit and experience nature as it exists in the Amazon, or the Arctic, or the iconic national parks of western North America.

As several of the chapters in this book demonstrate, urban rivers bisect some of our city's most marginalized and low-income neighborhoods. Urban rewilding has the potential to address the historical inequities regarding access to outdoor recreation and open space and provide residents with opportunities to increase their knowledge and appreciation of ecosystems. Certainly, bike paths and parks provide outdoor recreation, but this is fundamentally different from having a wetland or daylighted creek where kids can assess water quality and watch dragonfly predation in real time, perhaps even catch sight of a beaver or heron. In other words, if we do this right, we can revitalize our cities to make them more habitable for both humans and nature. But as the following chapters indicate, urban rewilding is rife with both challenges and opportunities.

Humans have a psychological need for wildness. As Henry David Thoreau noted, we need an infusion of arborvitae in our tea. We even have a new term—"nature deficit disorder" coined by Richard Louv, author of *Last Child in the Woods*—to describe the loss of contact with the natural world. Increased exposure of urban dwellers to biodiversity enhances our psychological well-being, fosters ecological literacy, and can even reorient our interior landscape to offset climate anxiety. The question is, how can we increase biodiversity, both for its own sake and human well-being in an increasingly urbanized world? Rewilding offers us opportunities to engage with the natural world—such engagement should be more akin to a conversation, rather than manipulation, moving us away from the hubris that has dominated much of our interactions with nature.

Human well-being goals, such as psychological health deriving from having green space, flowering plants, and birdsong, can also be met with non-native species. As ecologist Richard Hobbs writes, "Novel ecosystems may provide some of the most important opportunities to connect with nature for a wide cross-section of society [and] . . . could be critical to developing a greater sense of environmental responsibility or stewardship."[21] Some of our greatest ecological insights, such as those of renowned botanist Henry Cowles, have come from studying such urban and novel ecosystems.

Urban rivers allow us to rethink our adversarial relationship to the natural world. Much of our modern history has vacillated between egregious exploitation of nature to preservation of nature dedicated to an aesthetic ideal. By the mid-twentieth century, we began to recognize that nature had intrinsic rights and needs leading to species and ecosystem protection, and we saw ourselves as a scourge upon the planet. It seemed that everything we did just made things worse. The solution was to set strict limits on human behavior and imprison nature in national parks and reserves, leading to a further bifurcation of the nature-culture divide that has plagued western civilization for centuries and is at the core of our environmental crisis.

But if we begin to think of urban rivers as novel ecosystems where humans and nature interact in a holistic manner that prioritizes ecosystem function—with a variety of native and exotic species comingling in an ever-evolving dance to adapt to rapidly changing conditions—subsequent generations can experience nature as something that surrounds them, and humans become part of the natural system while nature becomes part of the human environment. The primary question is how exactly can our urban ecosystems sustain humans and other species? How can we grow the resiliency of urban ecosystems while simultaneously fostering an empathy and appreciation of a hybrid nature? As Aldo Leopold noted, "A thing is right when it tends to preserve the integrity, stability, and beauty of the biotic community. It is wrong when it tends otherwise." Perhaps thus, we should encourage nature's resilience, embracing the unplanned diversity, the accidental rewilding, so that we can develop a tough, new ecosystem capable of adapting to the massive changes we've wrought.

Ultimately, resilience, both cultural and ecological, might be a more meaningful goal for conservation than "naturalness," or restoration, to an unachievable standard, especially in urban environments. Ecological resilience is "the capacity of an ecosystem to respond to a perturbation or disturbance" by resisting damage and recovering quickly.[22] Resilience then implies a self-sustaining system, possibly self-willed and thus "wild."

NOTES

1. This is a generalization that largely applies to the white, Protestant, middle and upper classes. Certainly, America has long contained a multiplicity of cultural views of the natural world.
2. Schoenberg, *Gonzaga University*, 81.
3. Schoenberg, *Gonzaga University*, 100, 139.
4. "Gonzaga Obtains Fair 'Spectacular,'" *Spokane Daily Chronicle*, May 26, 1966.

5. Hobbs, Higgs, and Harris, "Novel Ecosystems," 599–605; Morse et al., "Novel Ecosystems in the Anthropocene," 12.
6. Hobbs, Higgs, and Hall, *Novel Ecosystems*; quote from Morse et al., "Novel Ecosystems in the Anthropocene," 12.
7. Richard Conniff, "Urban Nature: How to Foster Biodiversity in World's Cities," YaleEnvironment360, January 6, 2014, http://e360.yale.edu/feature/urban_nature_how_to_foster_biodiversity_in_worlds_cities/2725/.
8. Hobbs et al., "Managing the Whole Landscape," 557–64.
9. Hobbs, Higgs, and Hall, *Novel Ecosystems*, 310–25.
10. McKinney, "Urbanization, Biodiversity, and Conservation," 885–86.
11. Martin, *Wild by Design*, 10–11.
12. "Mission and Vision," Society for Ecological Restoration, accessed October 9, 2022, https://www.ser.org/page/MissionandVision.
13. "Cheonggyecheon Stream Restoration Project," Landscape Performance Series, accessed December 6, 2021, https://www.landscapeperformance.org/case-study-briefs/cheonggyecheon-stream-restoration; Andrew Revkin, "Peeling Back Pavement to Expose Watery Havens," *New York Times*, July 16, 2009.
14. "Los Angeles River Revitalization Master Plan 2007," LA City, accessed May 2, 2015, http://boe.lacity.org/lariverrmp/index.cfm.
15. Jackson, "Walking through Time."
16. Dearborn and Kark, "Motivations for Conserving Urban Biodiversity," 437; Marris, *Rambunctious Garden*, 119; Mascaro et al., "Origins of the Novel Ecosystems Concept," 49.
17. Marris, *Rambunctious Garden*, 108; Davis, *Invasion Biology*, 108.
18. "Trails Coalition Receives Hangman/Latah Creek Planning Grants," Spokane River Forum, December 14, 2016, https://spokaneriver.net/news/trails-coalition-receives-hangmanlatah-creek-planning-grants/. Unfortunately, this plan has yet to be implemented.
19. Adam Shanks, "Coeur d'Alene Tribe Buys Latah Creek Property, Spares It from Housing," *Spokesman-Review*, September 1, 2021.
20. "68% of the World Population Projected to Live in Urban Areas by 2050," UN Department of Economic and Social Affairs, accessed September 1, 2020, https://www.un.org/development/desa/en/news/population/2018-revision-of-world-urbanization-prospects.html.
21. Hobbs et al., "Managing the Whole Landscape," 557–64.
22. Walker and Holing, *Resilience Thinking*, 20.

BIBLIOGRAPHY

Davis, Mark. *Invasion Biology*. Oxford: Oxford University Press, 2009.

Dearborn, Donald C., and Salit Kark. "Motivations for Conserving Urban Biodiversity." *Conservation Biology* 24, no. 2 (April 2010): 432–40.

Folke, Carl, Stephen R. Carpenter, Brian Walker, Marten Scheffer, Terry Chapin, and Johan Rockström. "Resilience Thinking: Integrating Resilience, Adaptability and Transformability." *Ecology and Society* 15, no. 4 (December 2010): 20.

Hobbs, Richard J., Eric Higgs, Carol M. Hall, Peter Bridgewater, F. Stuart Chapin III, Erle C. Ellis, John J. Ewel, Lauren M. Hallett, James Harris, Kristen B. Hulvey, Stephen T. Jackson, Patricia L. Kennedy, Christoph Kueffer, Lori Lach, Trevor C. Lantz, Ariel E. Lugo, Joseph Mascaro, Stephen D. Murphy, Cara R. Nelson, Michael P. Perring, David M. Richardson, Timothy R. Seastedt, Rachel J. Standish, Brian M. Starzomski, Katherine N. Suding, Pedro M. Tognetti, Laith Yakob, and Laurie Yung. "Managing the Whole Landscape: Historical, Hybrid, and Novel Ecosystems." *Frontiers in Ecology* 12, no. 10 (2014): 557–64.

Hobbs, Richard J., Eric S. Higgs, and Carol M. Hall, eds. *Novel Ecosystems: Intervening in the New Ecological World Order*. Hoboken NJ: Wiley-Blackwell, 2013.

Hobbs, Richard J., Erik Higgs, and James A. Harris. "Novel Ecosystems: Implications for Conservation and Restoration." *Trends in Ecology and Evolution* 24, no. 11 (November 2009): 599–604.

Jackson, Stephen. "Walking through Time in the Lake Michigan Dunes." *Frontiers in Ecology and the Environment* 9, no. 527 (November 2011): 526–27.

Marris, Emma. *Rambunctious Garden: Saving Nature in a Post-Wild World*. New York: Bloomsbury, 2011.

Martin, Laura. *Wild by Design: The Rise of Ecological Restoration*. Cambridge: Harvard University Press, 2022.

Mascaro, Joseph, James A. Harris, Lori Lach, Allen Thompson, Michael P. Perring, David M. Richardson, Erle C. Ellis. "Origins of the Novel Ecosystems Concept." In *Novel Ecosystems: Intervening in the New Ecological World Order*, edited by Richard J. Hobbs, Eric S. Higgs, and Carol M. Hall, 45–57. Hoboken NJ: Wiley-Blackwell, 2013.

McKinney, Michael. "Urbanization, Biodiversity, and Conservation." *BioScience* 52, no. 10 (October 2002): 883–90.

Morse, Nathanial B., Paul A. Pellissier, Elisabeth N. Cianciola, Richard L. Brereton, Marleigh M. Sullivan, Nicholas K. Shonka, Tessa B. Wheeler, and William H. McDowell. "Novel Ecosystems in the Anthropocene: A Revision of the Novel Ecosystem Concept for Pragmatic Applications." *Ecology and Society* 19, no. 2 (June 2014): 12–22.

Schoenberg, Wilfred P. S. J. *Gonzaga University: Seventy-five Years 1887–1962*. Spokane: Gonzaga University, 1963.

Walker, Brian, and C. S. Holing. *Resilience Thinking*. Washington DC: Island Press, 2006.

Part 2 TRIBUTARIES

How City Rivers Make River Cities

Part 2 explores how reimagining urban waterways can change the identity of a city and what happens when a community views its river as an asset rather than as a liability. As these four chapters illustrate, the revival of urban rivers is primarily based upon bringing people to the river and the resulting changes in attitudes toward the river. Changes in both infrastructure and mindsets work hand in glove to foster rewilding.

In the first two selections, historians Char Miller and Jennifer Stevens detail how two cities—San Antonio and Boise, respectively—have embraced their namesake rivers so much that each city's identity is now tied to that of its river. In these cases, not only is the physical river a hybrid of human infrastructure and natural forces, but our mental map of the city and river become inexorably bound as well. The next two chapters examine rivers lost and reclaimed—how Milwaukee and the Salt Lake Valley are rediscovering their waterways and reimaging their relationships to their rivers.

Drawing on his extensive research on the environmental history of San Antonio, Char Miller, in "Wild, Managed, and Reclaimed," captures the city's early recognition of its river as an amenity rather than a threat. The particular geology and hydrology of the Edwards Plateau led to intense flooding of the San Antonio Valley, yet at the same time, residents "began to reimagine the river not as a source of danger but of commerce." Even at the height of frontier expansion, visitors, residents, and entrepreneurs recognized the river as integral to the city's identity. In the twentieth century, the river became a stage on which various narratives could be superimposed, evolving into the world-famous River Walk, where human engineering and imagination combined with natural systems to create a hybrid riverscape, but one that is constantly shifting on the winds of human imagination.

San Antonio might have been the first American city to substantially rethink its relationship with its river. Less well-known is how the San Antonio River Authority and the city's marginalized neighborhoods also embraced river revitalization. In many ways, San Antonio is the story of two cities and one river. At the same time as the city's River Walk became an international attraction and an economic mainstay, low-income Latino neighborhoods continued to experience flooding, provoking the standard engineering response of channelization and concrete, which resulted in the creeks becoming ugly and neglected. But then these communities began to reimagine their creeks, and by 2020 the inner-city waterways have transformed into a linear parkway with eighty miles of trails.

While San Antonio embraced its river to forge an identity based on the mythology of the colonial Southwest, Boise, Idaho, in another frontier tradition of western settlement, obliterated its past to create a new identity. Boise's near total deconstruction of its downtown allowed for a complete reimagining of the role of the river in the lives of its citizens, and the city became the quintessential representation of the New West, with a high-tech economy built around access to outdoor recreation. But as Jennifer Stevens demonstrates in "From Floodplain to Kayak Park," this process was far from intentional and was almost accidental. Stevens unpacks how a dammed and highly regulated river became the focal identity for one of the fastest growing cities in the United States. While it could have unfolded in an entirely different manner, Stevens highlights a series of unforeseen contingencies that led to the creation of one of the nation's best examples of urban rewilding with the establishment of the twenty-one-mile Boise River Greenbelt.

While San Antonio and Boise remade themselves foregrounding their rivers, Milwaukee and greater Salt Lake City are currently engaged in a process of revitalization to reconnect people to their respective urban waterways. In San Antonio, well-heeled visitors stroll, shop, and dine along the river, and in Boise, white (mostly) middle-class residents see the river as part of a recreational lifestyle, while Milwaukee and Utah work to refashion their waterways into opportunities for more marginalized communities to connect to the natural world in their own neighborhoods.

As Christian Young in "Urban Ecology" demonstrates, public policy is not the only pathway for urban rewilding. With a focus on Milwaukee's Urban Ecology Center, Young shows how environmental education centers and the need for outdoor classrooms can initiate urban rewilding and foster environmental stewardship—a bottom-up approach. Also following a deliberate process to reclaim its waterways, Milwaukee worked to alleviate the scars of its industrial past despite the constraints of old infrastructure. Recovering a lost river can happen in multiple ways, not just through polices; it can also occur through physical engagement with the river.

Marian Rice, in "The Lost River," provides a case study of successful river revitalization on Utah's Jordan River. Rice details the fits and starts of creating a governing structure to address the challenges of managing a waterway that crosses multiple municipalities. The Jordan River reveals the complexity of the political challenges posed by urban river management and how the health of the river itself can serve as the touchstone for overcoming these challenges. In this case, a city's identification with a river doesn't just happen but needs to be deliberately cultivated.

As all these chapters illustrate, the process of reimagining a river coincides with the physical revitalization in a feedback loop that becomes rewilding. As citizens, governments, and nonprofits roll up their collective sleeves and clean up the industrial waste and engage in replanting, artwork, and river celebrations, making the river a more inviting place for recreation and contemplation, the policy formation begins to take on a more imaginative dimension of what is possible. Rewilding becomes the nexus between action (revitalization) and imagination.

4 Wild, Managed, and Reclaimed

An Environmental History of the San Antonio River Watershed

Char Miller

At the December 11, 1926, dedication of the Olmos Dam—a 1,900-foot, gleaming-white concrete structure that spans the Olmos Valley just north of San Antonio, Texas—the city celebrated. Its mayor, John Tobin, vowed that the dam would forever protect the downstream community from future inundations that had plagued it since its eighteenth-century beginnings. "San Antonio's old flood nuisance is now ended," Tobin told the assembled throng, "and there is nothing to stand in the way of her progress and growth." A significant sign of the community's emerging modernity, the Olmos Dam was "the greatest step that San Antonio has taken towards becoming the leading city of the Southwest."[1]

For all its size and solidity—the dam contained some 418,000 pounds of steel and 90,000 cubic yards of concrete—and for all its projected innovative impact, the Olmos Dam was problematic in two important respects. Local and national engineers who inspected the completed facility reported to the city that unlike any other contemporary flood-retention structures with which they were familiar, the Olmos Dam's sluice gates were flawed. Built at ground level, they predicted that flood debris would quickly clog the gates, rendering them ineffective. The other worrisome feature was that the gates had to be operated manually, unlike similar projects that were machine-driven. The engineers vigorously criticized these structural elements and believed that they weakened the dam's functionality, perhaps cripplingly so.[2]

Although the dam did not fail, one curious consequence of its inexpert engineering was that the gatehouse in which the sluice controls were situated had to be continuously monitored. With a lot of time on their hands, the operators began to keep a record of the floodpool elevations—the height of the water that backs up against a dam's impounding face. They established the structure's base elevation of 678.34 feet above sea level, and beginning in 1936, they measured each flood event—potential or realized. Some thunderous storms barely registered at all: heavy rains on April 26, 1941, recorded a mere six-inch rise behind the dam. In March and early September 1946, rainfall that submerged other parts of the city outside the dam's protective zone did not rise even an inch behind the dam. The operators may have expected the same minimal result at 2:00 a.m. on September 27, 1946: the measurement stood at zero feet. Ten hours later, the Olmos Dam held back a whopping thirty-eight feet of water, the highest ever recorded, then or since.[3]

The operators' informal recordkeeping was handwritten on the walls of the gatehouse and yet spoke formally about the dam's significance. The data they collected gave floods scale and legibility, graphic representations of the critical role that the Olmos Dam played in controlling stormwaters upstream so that the central core of San Antonio remained higher and drier than it had in the past. The dam's pivotal placement, seen from the gatehouse or from car windows rolling past Olmos Drive across from the dam, tells two stories. Upstream, and to the north of the dam, the river and its main tributary, Olmos Creek, was uncontrolled until much later in the twentieth century; every time a heavy storm dumped inches of rain over the upper reaches of the watershed, stormwaters roared downstream, as wild and rampaging as ever. Downstream, and to the dam's south, however, the San Antonio River's central trunk was far more placid. Indeed, since the dam's mid-1920s construction, and with the goal of moving water through the city as quickly and safely as possible, the river's former serpentine course was straightened, its banks concretized, bridges elevated, and once river-encroaching structures—buildings, flour mills, and check dams—eliminated. That was what Mayor Tobin had in mind when he asserted that the Olmos Dam ensured ultimate control of the river and of its episodic power over the human experience of this once-dangerous stream.

Tobin was only partly correct. In truth, there were two rivers in one: a hybridized riparian system, managed and unmanaged. A hybridity that historian Richard White, in reference to the Columbia River, has likened to an "organic machine" in which a waterway doubles as "a natural space and a social space."[4] This perception of a river as a "coupled natural and sociocultural system," argues landscape architect Thaïsa Way, facilitates the ongoing process of "reimagining any future for both river cities and city rivers."[5] Her point is made manifest in San Antonio. The San Antonio River flood-control system would not be fully locked in place until the 1990s, a lag that was particularly evident on the flood-prone west side creeks that flowed through the poorer, Latino-majority neighborhoods. Yet even before the river and creeks were fully concretized, some in the community had begun to rethink their relationship to those segments of the local watershed that had become hardened, inaccessible, forbidding. Once again, the river would be reconceived, a reclaiming and recreating that closely integrated the natural and human environments, which, in the process, increased riparian, environmental, and public health.

Flash-Flood City

The San Antonio River, which rises out of the Edwards Plateau in south-central Texas, is one of the most tightly managed riparian ecosystems in Texas. The reason for its highly structured character—dams, channels, tunnels, and other infrastructure—is that a slashing rain can turn this relatively tranquil watercourse into a rampaging torrent. So potent are these stormwaters that "Flash-Flood Alley" is how the National Weather Service describes the San Antonio River and other streams that flow out of the plateau, cut through the Balcones Escarpment, and press south-westward toward the Gulf of Mexico. The communities most imperiled, then, hug the Balcones Escarpment, a geological fault zone that runs for roughly 450 miles. Like a lopsided smile, it curves east and north from Del Rio on the Rio Grande all the way to the Red River delineating the Texas-Oklahoma border; its hilly contours define the cities of San Antonio, New Braunfels, San Marcos, and Austin, and from Austin north to Georgetown and beyond. (A modern signifier is Interstate 35, which from San Antonio north parallels the escarpment to its east.)[6]

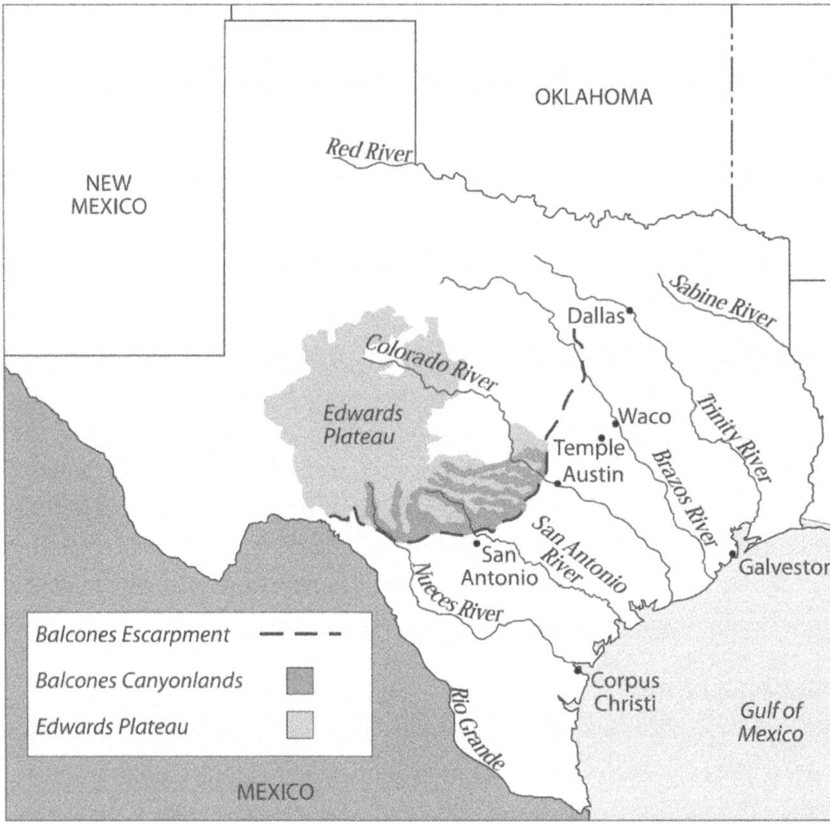

2. Map of the Edwards Plateau, Balcones Escarpment, and river systems. Created by Charles David Grear, used by permission.

This rumpled and craggy landform demarcates the southern terminus of the Great Plains, an elevated terrain that in Texas is known as the Hill Country; some sections of the rough limestone and thin-soiled landscape rise to an elevation of two thousand feet or so. The land that slopes away from the fault line toward the Gulf of Mexico is the southern coastal plain. The reciprocal relationship between these two masses and long-standing climatological patterns can produce wild swings in local weather. Because this region falls loosely within the transition zone between the humid eastern section of the United States and the arid west, the climate can toggle between deluge and drought, an oscillation fueled in part by whether an El Niño or La Niña weather system prevails. Another trigger mechanism involves the Gulf of Mexico. When warm, moisture-laden air pushes onshore, the flow slowly lifting with the rise in elevation. Once it reaches the escarpment, the uplift is more abrupt, forcing the moist air to interact with colder temperatures above. The moisture condenses, the cooler air falls, then it is warmed and rises again, a cycle of convection that can result in major thunderstorms. "It has long been recognized," noted C. Terrell Bartlett in a contemporary assessment of the cause of the 1921 flood, "that in many cases the sudden rise along the Balcones Escarpment causes intense precipitation along and just above its margin."[7]

Over the millennia, rains have carved into the soft limestone a series of creeks, streams, and rivers that widen as they reach the coastal plain and head to the gulf, an erosive process that flooding could accelerate. In their more placid states, the San Antonio River, the Guadalupe River (that flows through the city of New Braunfels), the Comal River (that San Marcos straddles), and the Colorado River (that bisects the state's capital, Austin) sustained generations of Indigenous communities; in time Spanish, Mexican, and American settler-colonists occupied these same sites. Each group took advantage of these systems' life-sustaining properties and the ecological abundance that came from living within the fertile intersection of different biozones—prairie and plateau, grassland, riverine, and woodland. But should a furious thunderstorm explode over the upper reaches of the local watersheds, the floodwaters that sluiced down innumerable gullies and ravines and then surged into streams could have devastating consequences. Within moments, the almost bone-dry San Antonio River and its arroyo-like tributaries could become churning torrents.

That is what happened on September 9–10, 1921: San Antonio went under water. So did New Braunfels, San Marcos, and Austin, along with the smaller communities of Taylor and Thrall. The staggering volume of water that came crashing down was the result of a slow-moving tropical depression that several days earlier had crashed ashore in northern Mexico. As it swung over the Rio Grande, the system dumped upwards of 6 inches on Laredo, inundating low-lying neighborhoods. Pressing north, it cycled along the Balcones Escarpment, where storm cells unleashed their full fury. Thrall recorded a record-setting 38.21 inches in a twenty-four-hour period, believed to be one of the largest, single-day rainfall events in the continental United States. More than 23 inches fell on Taylor and Austin received 18.23.[8]

These fixed-site numbers do not tell the whole story, because upstream of each of these communities, heavy rains were coming down over the folds of the Edwards Plateau and surging into the region's major and minor river systems. These roiling waters swept over farms, ranches, and feedlots; houses and barns were pushed off their foundations and careened downstream; uprooted trees became battering rams, slamming into buildings, tearing up transportation infrastructure, and collapsing bridges. Property damage was severe, but the number of fatalities was even more so: in Taylor eighty-seven people died, and another six perished in its home county of Williamson. Six people drowned in Travis County, where Austin is located. By all measures, the 1921 flood was the most devastating in the history of the Lone Star State.[9]

San Antonio's suffering was linked to its eighteenth-century Spanish design. Its planners had platted the frontier town in a floodplain: its streets, plazas, and residential areas lay within the embrace of two waterways, the San Antonio River on the east and San Pedro Creek to the west (their confluence lay downstream of the city's central core). As a result, the community periodically had been inundated. Until its roads were hardened in the late nineteenth century, even light rains had turned streets into a boot-sucking morass. The foundations of new upper- and middle-class housing built after the mid-nineteenth century often were lifted above street level in hopes of keeping their occupants clear of moderate flooding. But nothing protected the downtown sector during the floods of 1819, 1845, and 1865 when so much water poured downriver and through the creek that their combined overflow merged, trapping residents and devastating homes, churches, and commercial properties. Two punishing floods in 1913 and 1914 shredded parts of the city, especially its impoverished west side. These floods added to the pattern of inundation that was built into San Antonio's environmental history. Between the megafloods of July 1819 and September 1921, at least fifteen significant floods submerged some or much of the community.

Repeated flooding made and remade the San Antonio region over time. Their impact can be read in the contours of the shape-shifting structure of the river and valley in which the city is located, the composition of its

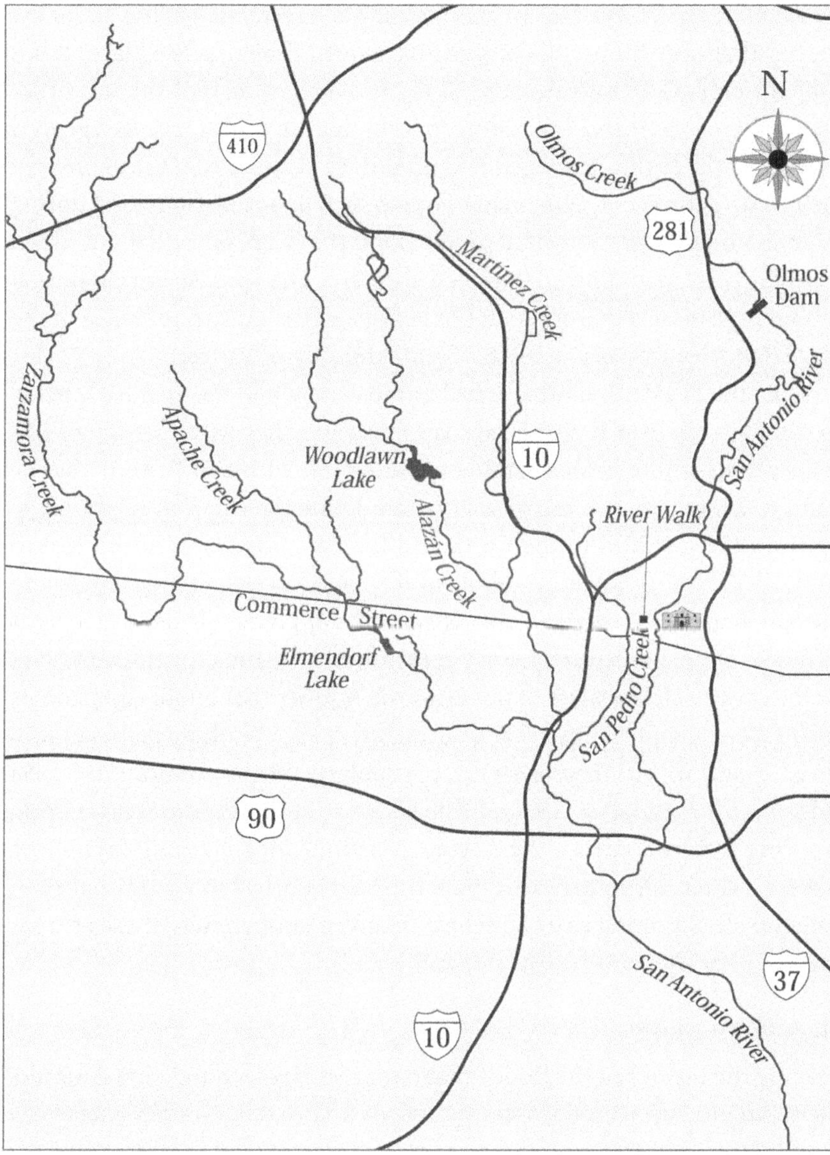

3. Map of San Antonio River and its tributaries. Created by Charles David Grear, used by permission.

soils, the health of its riparian habitats, and the resilience of the countless species that over thousands of years have been drawn to the place. It has been a rich land in part because of the floods that carried silt and other material down to the valley adding to its biological health. The Indigenous peoples of the region managed and used these resources in the vicinity of their semi-permanent settlements close to springs and streams, and during their seasonal migrations that arced along the waterways that cut down from the plateau and through the escarpment. The Spanish entradas took careful note of the environmental riches of and the human presence in the place that they would name after Saint Anthony of Padua.[10]

What the Payaya and other tribal groups knew, and the Spanish quickly would discover, was that the San Antonio valley lies within a web of rivers, streams, and creeks. Imagine a handprint, with San Antonio as the palm. Like fingers, five major waterways define the riparian relationship between the low hills to the north and the flatlands below. From east to west, these systems include Cibolo and Salado Creeks, Olmos Creek and the San Antonio River (now described as the upper and lower watersheds of the San Antonio River), Leon Creek, and the Medina River. Each is fed by the crystal-clear waters of the Edwards Aquifer that bubble up from its springs, bogs, and wallows; each ultimately merges with the San Antonio River. Inlaid within these systems, particularly on the community's west side, is a set of smaller streams. The shortest of these, San Pedro Creek, is spring fed; by contrast, the Alazán, Martínez, Apache, and Zarzamora Creeks, which ultimately feed into the San Pedro, carry minimal flow, generally functioning as dry ditches. Yet given enough rain, these arroyos can rise quickly, powerfully, and fatally.[11]

High Water Marks

Tracing the history of the floods that have regularly scoured San Antonio, then, is bound up with the community's efforts—at first merely rhetorical, then halting and incomplete—to control these punishing waters and the death and damage they routinely wrought. Full control, or what residents and officials believed would be complete mastery of the local watershed, would not happen until the latter half of the twentieth century. Yet even before the Olmos Dam and other flood-control infrastructure was set in place, civic-minded advocates began to reimagine the river not as a source of danger but of commerce, a touristic space that would boom the local economy.

They were not the first to find the local river beguiling. The river's cool, spring-fed waters thrilled early eighteenth-century Spanish explorers, including Capt. Domingo Ramón, who wrote in his journal on May 14, 1716, "On this day I marched to the northeast seven leagues through mesquite bush with plenty of pasturage. Crossing two dry creeks we reached a water spring on level land, which we named San Pedro. There was sufficient water here for a city of one-quarter league."[12] Frederick Law Olmsted, who visited San Antonio in the 1850s, was also smitten by the river's sweep through the town: "We irresistibly stop to examine it and are struck with its beauty. It is of a rich blue and pure as crystal, flowing rapidly but noiselessly over pebbles and between reedy banks. One could lean for hours over the bridge-rail."[13] What mesmerized late nineteenth-century journalist and novelist Harriet Spofford was the river's serpentine course:

> Twisting and turning and doubling on itself, so tortuous that the three miles of the straight line from its head to the market-place it makes only in fourteen miles of caprices and surprises, rapids and eddies and falls and narrow curves, reach after reach of soft green and flickering sunshine, each more exquisitely beautiful than the other. Around every lane it takes a loop; here it is just a pebbly ford, there, although so perfectly transparent that you can see every flint in its bed, it is of a profound depth, and everywhere it is of a color whose loveliness is past belief.[14]

Spofford had arrived in the city in 1877 as a passenger on the inaugural trip of the Galveston, Harrisburg, and San Antonio Railroad, the first train to reach the south Texas city. Because she was a friend of the company's principal investor, Thomas W. Pierce, her enamored report may have been designed to encourage tourists to ride the rails to visit the quaint town through which the languid stream flowed. For his part, Pierce was convinced that tourism would bolster the city's economy. So was *Hanford's Register*, which touted the now-connected city's many virtues: "The completion of this road to the ancient city of San Antonio, world famous for its ancient and quaint-looking Spanish buildings; its 'Alamo' and Missions, and its beautiful serpentine river, with its crystal waters;

the mildness of its Winters, and renowned in having a climate which in itself has cured thousands affected with consumption and pulmonary complaints, opens to the tourist and emigrant of the finest portions of the Empire State to Texas."[15]

As much as he would benefit from that very migration, railroad executive Pierce worried that the iron horse might run roughshod over San Antonio's quaintness. He urged local politicians to confine modern growth to the town's periphery to protect its charming river. Local preservationists shared his concern, leading the *San Antonio Daily Express* to editorialize that the river was vital to the community's affective and aesthetic sense of itself. "Our river can best serve the interests of the city and contribute to its welfare and happiness as that of a clear beautiful winding stream through its center. Such a stream, like a silver thread running under bridges, darting around curves and sparkling in the sunlight, would make up for much that is lacking in the improvement and adornment of our city . . . its banks could be converted into flower beds, and pleasure boats [could] afford recreation . . . and our river would become the jewel of Texas."[16]

Yet, increased visitation could also tarnish the river's—and by extension, the city's—very gem-like quality, as revealed in the integration of the San Antonio River's sensual appeal and meditative character into the bourgeoning tourist economy. That economy took off in tandem with the city's explosive growth in the late nineteenth and early twentieth centuries: its physical expansion swallowed up farmlands that Spanish ditches once irrigated, and residents turned the river into a garbage dump. Its once-bounteous flow was further diminished as the local water company drilled artesian wells to slake the citizenry's growing thirst. The impact of this new extractive technology on the San Antonio River and San Pedro Creek was immediate; springs that fed them began to dry up. "I have seen this bold, bubbling, laughing river dwindle and fade away," George W. Brackenridge wrote a friend. "This river is my child and it is dying, and I cannot stay here to see its last gasps. . . . I must go." His maudlin projection was a direct consequence of his prior actions. Brackenridge owned the company that supplied the city with its potable water, and he drilled the first artesian well out of which gushed more than three million gallons a day. From his lavish family home at the river's headwaters, he daily observed how quickly that prolific well and others drained the river.[17]

By 1911 the *San Antonio Express* worried that the damage to the city's greatest "natural asset," its "winding tree-shaded stream" whose "sinuous course through the city . . . elicits the admiration of visitors," was so consequential that fewer tourists would travel to revel in its beauty: "The stream has dwindled to a sluggish current running through neglected banks over a riverbed covered with slime and silt."[18]

Sick of its stench, Anglo downtown business owners proposed that the river be piped underground, an argument that gained adherents following the devastating September 1921 flood that killed an estimated eighty people. Engineers drew up plans to dam the river's upstream flow, straighten and widen sections of its middle reach, and then force it into culverts just north of downtown. River reformers, also members of the Anglo elite, most notably the women who had established and led the formidable San Antonio Conservation Society, fought against the attempt to bury the river.[19] Their counter was that beautifying the river's banks would pay off, luring a carefree clientele to its waters: "At night myriads of electric bulbs will shine from the trees while Mexicans dressed in the garb of Aztec Indians will paddle canoes, filled with tourists."[20] This lurid settler-colonial and racist fantasy of the river as an entertainment zone intensified the process by which this natural space became theatrical, and even though the particular sets of this stage would change over time, the river remained an urban backdrop and dramatic foreground.

In 1929, for instance, the San Antonio Conservation Society sponsored Robert H. H. Hugman's redesign of the river's Great Bend: as the south-flowing river entered the central core, it broke east, then south, then west, before swinging south once more, creating a mile-long loop and a peninsula-like wedge of land. It was this bend's banks that Hugman planned to beautify. Fascinated by the vibrant French Quarter in New Orleans, where he had worked as an architect in the mid-1920s, Hugman concocted a clever scheme that incorporated flood-control features so as to protect the refurbished landscape. This plan included the creation of a bypass channel that would flush floodwaters directly south past the Great Bend, with gates that could be raised to keep rampaging floods out of the refashioned terrain. New limestone pathways would be laid down and bridges would arch over the river, and the whole would be dotted with stores, hotels, and restaurants.[21]

Named the Shops of Aragon and Romula, Hugman's project wrapped a mundane commercial enterprise in a patina of Spanish exoticism, a romanticized riverscape in which visitors would frolic and consume. "We read of descriptions of the old cities of Spain," he wrote, "of a narrow winding street barred to vehicular traffic yet holding the best shops and cafes; prosperous yet alluring, with its shadowed doorways and quaint atmosphere." His innovation was to reimagine that streetscape as a riverscape. Imagine boating "down the river on a balmy night fanned by a gentle breeze carrying the delightful aroma of honeysuckle and sweet olive," Hugman enthused. "Old fashioned street lamps casting fantastic shadows on the surface of the water; strains of soft music in the air; all of this would be the night life of Romula."[22]

Some of Hugman's ideas were realized in the late 1930s. The support of the mayor, Maury Maverick, and an infusion of Works Project Administration dollars and city investment produced seventeen thousand linear feet of walkways, fanciful stairs, woven-rope ladders, quaint bridges, and extensive landscaping. Adding to the area's allure was the construction of the open-air Arneson River Theater and the restoration of nearby La Villita Historic Arts Village, site of one of the city's earliest neighborhoods.[23]

What evolved into the modern River Walk—a mile-long pedestrian pathway set with the river's Great Bend in the downtown core—gained a major boost as a result of post–World War II prosperity. Greater disposable income and increased access to automotive mobility brought thousands of visitors to the Alamo City. Banking on these developments, a coalition of business and civic leaders, along with politicians and urban planners, organized HemisFair '68, a world's fair-like extravaganza. Tapping federal, state, and local coffers, they converted a large portion of the central business district and a nearby historic neighborhood into a visitors' playground and extended the river by constructing a concrete lagoon in front of the new convention center. Proponents argued that the enhanced river would stitch together the disparate elements of this new urban fabric. A prime example of this reknitting was Palacio del Rio, a twenty-one-story Hilton Hotel; among its innovative features were street- and river-level entrances that funneled guests and passersby between restaurants, shops, and convention facilities.[24]

Since then, most riverfront hotels in San Antonio have copied this pattern, generating substantial foot traffic—fourteen million visitors annually stroll down the contemporary River Walk's three miles of graceful, tree-lined paths, making it one of the great pedestrian experiences in the United States. Their spending—an estimated $3 billion annually—has become key to the River Walk's profound economic impact.[25] This now-historic pedestrian landscape also serves as the communal artery and civic stage, albeit without buff Aztec warriors plying its languid waters. Each year its meandering walkways and calm surface become home to innumerable processions and parades, celebratory moments that reflect a diversity of cultural, political, social, and religious values. Even more spontaneous celebrations sweep down the river: whenever the San Antonio Spurs win the NBA championship, the victorious team floats on motorized barges past hundreds of thousands of cheering fans.[26]

To expand the River Walk's value, in May 2008 local boosters and political leaders successfully encouraged voters to support the continuation of a venue sales tax that would underwrite $125 million worth of improvements to the River Walk, extending it north and south from the downtown core. In May 2009 the northern, or "museum reach of the river," a 1.3-mile-long greenbelt, opened with great fanfare; it links the downtown with several cultural institutions, among them the San Antonio Museum of Art and the Pearl Brewery, which has been rehabbed into a work-live-and-play zone. By 2013 the southern stretch, or "mission" reach, more directly connected the city with the four historic missions forming the San Antonio Missions National Historic Park—Concepción, San Jose, San Juan, and Espada.[27]

These expansions have altered the natural river's form and function, if in different ways. The northern segment, for example, contains a series of locks that integrates the river with an already highly engineered system artfully concealing flood-control tunnels, pumps, and water-recycling infrastructure; barges now ply these placid waters. By contrast, the mission reach was paired with an extensive and multi-million-dollar riparian restoration project that the U.S. Army Corps of Engineers and the San Antonio River Authority managed. It was designed to return some of the lower river to its original sinuous course, repair damaged wetlands, restore riparian habitat, and replant native trees and other vegetation.

The goal was to soften the formerly off-putting hardscape, to break up the concrete to allow the land to breathe, and to give pedestrians access to what had been long inaccessible. "Understanding San Antonio as a river city requires more than just highlighting a narrow channel of water that passes through an urban core," reminds landscape architect David Madla. "The river has been a site of continued innovation that drives the evolution of the city."[28]

Fingers of Green

As innovative as the redevelopment of the central trunk of the San Antonio River has been, in truth, it essentially reinforced the city's longstanding commitment to riverside tourism that would serve the city's many visitors. By contrast, the transformation of the interwoven west side creeks was dedicated to and for local residents, to those living in neighborhoods that the 1921 flood had ravaged and that until the late twentieth century were left unprotected from surging stormwaters. One of these inundations, which occurred along Zarzamora Creek in August 1974, might have just been another in a long line of floods that chased people from their homes, tore up the streetscape, and waterlogged businesses. Instead, it sparked a Latino uprising led by a new grassroots organization—parish-based, women-led, coalition-driven—known as COPS (Communities Organized for Public Services). The first public service its activists demanded was flood control. In a series of raucous meetings with the city council, they revealed a pattern of neglect that was responsible for the death and destruction that so frequently had damaged their community. Their research into the city's archives, for example, revealed that local bond money that had been set aside to reconstruct the west side creeks was never spent, a disturbing indicator of the city's racist politics that COPS intended to disrupt—and did. Within months a chagrined political leadership had committed substantial funding amounting to millions of dollars to channelize Zarzamora Creek, and in collaboration with the U.S. Army Corps of Engineers, to deepen, widen, and harden portions of the Alazán, Martínez, and Apache Creeks. Much of this new construction and related infrastructural upgrades across the west side were completed by the early 1990s. Activism by COPS had largely resolved this key environmental injustice that had long plagued San Antonio's west side, Latino-majority neighborhoods.[29]

However essential, this protection came with an unanticipated cost. The broader, deeper, hard-walled creeks and their expanded, fenced-off banks cut through the community like large scars. Neither aesthetic nor accessible, these channels had become visual barriers and for some sparked a new level of trepidation: infrequently maintained, the creeks filled with trash and other detritus, and seemed to offer havens for illicit activities. The infrastructure may have controlled rampaging stormwater, but it also brought a new set of anxieties and hazards. As a child, Robert Ramirez, who grew up on the west side near the Zarzamora and Apache Creeks, was repeatedly warned not to explore along the streams; they were too dangerous of a place to play. Diane Williams heard the same cautionary words about Salado Creek, which arcs along the city to the northeast. As for Irene Moreno, she told a reporter that she was "scared" to visit Zarzamora Creek, a community-wide fear: "Nobody came through here."[30]

Note Moreno's use of the past tense. Her conversation with the local journalist occurred at the 2010 dedication ceremony of the newly opened, greened-up stretch of the Zarzamora. That November Julián Castro (the mayor), David Medina (councilman), Rod Radle (of the Metropolitan Alternative Housing Corporation), and Robert Ramirez (of the Westside Creeks Restoration Project [WCRP] oversight committee) cut a long, red ribbon to mark the official opening of the Apache Creek Linear Park. The first greenway on the city's west side, it follows the two creeks from General McMullen Drive in a southeastern direction to West Commerce Street where the Zarzamora enters Lake Elmendorf. The fully landscaped, bench-lined asphalt path serves walkers, joggers, and cyclists. These recreational opportunities were not the only assets that captured the attention of those attending the dedication. Changed, too, Moreno noted, was her perception of the formerly off-putting creek. As cormorants and egrets winged by, she meditated, "It feels real calm, like you can come out, sit and relax." Shouted enthusiastic walker Jeronimo Gueda, "I can't believe we are on the West Side!"[31]

In the ensuing decade, that same enthusiasm spread with each new mile of linear parks that unfurled along Alazán, Apache, Martínez, San Pedro, and Zarzamora Creeks; each linked with the others into an extensive network of pedestrian paths. The concept had been a long time in the making. In the early 1980s, urban planner Ralph Bender had developed the idea of a city-wide series of linear parks, a system he called "Fingers of Green."[32] By the 1990s the city's Open Space Advisory Board was pushing for its implementation, as was zoning commissioner, later city council representative, Howard Peak. When Peak became mayor of San Antonio in 1997, he used his position to advocate for local funding of the project, and after his tenure, he continued to press for its construction. His unstinting support brought this honor: the Howard W. Peak Greenway Trails System now bears his name.[33]

Expediting the reclamation of the west side creeks was the result of the on-the-ground management that the WCRP oversight committee provided. Within a decade the WCRP had conducted community-based research, held public charettes and policy-planning sessions, and in collaboration with the city of San Antonio and the San Antonio River Authority, helped give this nested system of waterways, once so terrifying because their flood-scouring energy, a more welcoming mien.[34] One example of many was the second facelift that the Zarzamora received. In 2019 the San Antonio River Authority spent tens of thousands of dollars relandscaping another stretch of the formerly flood-prone creek; mountain laurels and sycamores, cedar elms, oaks, and crabapples, native shrubs and grasses softened the hardscape and beautified what had been so ugly.[35] By 2020 more than $300 million had been invested in restoring local waterways and the greenway linear park system, turning them into a recreational hub that enables walkers and bikers to connect with eighty miles of trails paralleling the inner-city creeks.

The restoration of the west side creeks is also woven into a landscape-scale set of initiatives to reengage with, even renaturalize, the region's larger watershed. The handprint analogy—with San Antonio as the palm and the five major rivers and creeks as the fingers—helps to explain the scope of this reengagement. Starting where these riparian systems and their many tributaries begin, water cuts through the Balcones Escarpment or bubbles up from springs, bogs, and seeps before running southeast on the way to the Gulf of Mexico. In the northern reaches of Bexar County, along the upper extent of Salado, Olmos, and Leon Creeks, among others, check dams have been slotted into the thin soil to slow down floodwaters. Those structures that overlie the Edwards Aquifer recharge zone serve a crucial second purpose: to percolate water into the vast karst aquifer that is the key water supply for the city of San Antonio. Even as activists demanded this innovative solution, they collaborated with likeminded groups across the city and with sympathetic public officials to place on the ballot a series of parks and aquifer-protection propositions. The propositions gained broad support and have funded aquifer protection, restored creek bed and creek bank, renewed aquatic and riparian habitat, and greened the now ubiquitous linear parks.[36]

If there is a silver lining in the city's complicated history of flood control and its initial fixation on such major interventions as the Olmos Dam and associated concrete-and-steel infrastructure, it is that as it slowly rolled out these protective measures, some residents, activists, planners, and public officials, along with local, state, and federal agencies, began to reimagine the community's relationship with its sprawling watershed. Emerging from these reflections was a more beneficent perception of the many creeks, streams, and rivers that give shape to local topography and geography. In time, these agents of change would demand an equitable distribution of flood-control infrastructure and later still dream up a series of accessible linear parks and an accompanying network of recreational amenities that have drawn people back to the river and its tributaries that, until the late twentieth century, residents understandably feared.

One pivotal feature of this recreated relationship with the watershed is embedded in Confluence Park, built on a wedge of land just north of where San Pedro Creek empties into the San Antonio River. During the 1921 flood, their combined and roiling waters generated a maelstrom-like force. The 3.2-acre site today has a more pacific purpose: to educate the community about the river system, starting with its upland catchment basin down its 240-mile run to the Gulf of Mexico. Its visitors—especially the more than thirty thousand local school children who have toured the park since its 2018 opening—also learn about the ecology, hydrology, and history of local creeks and streams. These lessons are imparted in its classrooms, through observation and field studies in its native-plant terrain, and through the low-impact architecture that conveys the value of passive cooling and heating, solar-powered illumination, rain capturing, green roofs, bioswales, and other infrastructure that slow the movement of water across the site before it drains into the creek and river.[37] At this "living laboratory," the community receives hands-on experience of what it means to live within a watershed.[38]

However diminished the fear that once dominated perceptions of the San Antonio River and its floodplain, and without denying that this tangled riparian system is more ecologically resilient or that the ever-expanding greenway can and has enhanced public health, it would be a mistake to conclude that the river is no longer a river or that the watershed has become dormant. In 1998 and again in 2002, major storms exploded over the upper reaches of the San Antonio River and its tributaries, flushing so much water downstream that when impounded behind the Olmos Dam, U.S. 281, the city's central north-south highway, was submerged for several days. On the west and north sides, where roads dipped into creeks—called low-water crossings—roaring stormwaters swept away cars and damaged infrastructure. In some places, even the wider west side creeks overtopped their banks, spilling into adjacent neighborhoods.[39] In response, Bexar County and the city of San Antonio began using local funds to purchase from willing sellers more than three thousand houses and businesses located in the floodplains that these two flood events had identified.[40]

Other flood-wary lessons came in 2015. While San Antonio escaped the brunt of that year's Memorial Weekend floods that ravaged communities to its immediate north, hydrologists at the University of Texas at San Antonio calculated that the rapid increase in impervious urban development, including rooftops, roads, and other paved surfaces, and the corresponding loss of absorbing ground cover—all features consistent with San Antonio's sprawled pattern of growth—had accelerated those floods and their fatal power. Local media used these scientific findings to warn against complacency.[41]

Hurricane Harvey sounded its own tocsin. The 2017 superstorm, which stalled over Harris County (Houston), dropped upwards of fifty inches of rain there, generated river and flash flooding throughout the metropolitan area, and killed thirty-seven people. Swift-water rescuers saved more than ten thousand people and an estimated thirty thousand found temporary shelter in evacuation centers. Once again, the effects of a sprawled urbanism were manifest. Researchers noted how impervious infrastructure sped up the movement of the drenching rains once they hit asphalt and concrete. They also pinpointed the role that this material played in magnifying extreme precipitation, that is, the "urban effects on the organized tropical rainfall of a hurricane, in particular during one like Hurricane

Harvey, which stalled for several days."[42] More rain fell, in short, because of urbanization.

Planners in San Antonio took careful note of what had happened in Houston. In November 2017 the San Antonio River Authority (SARA) produced maps based on Federal Emergency Management Agency (FEMA) modeling that showed what a Harvey-sized storm might do locally. The news was grim: the resulting precipitation would overtop the Olmos Dam and downstream flooding would be devastating, particularly given how much development has occurred south of the dam since its completion in 1926. The impact would be much worse than SARA's images suggested because inexplicably FEMA's model did not incorporate street flooding (only riverine). That meant that the data set that SARA utilized produced flooding only within a narrow, two-mile corridor south of the dam and thereby missed all of the river's catchment basin, the various tributaries that course through the city's eastern and western sectors, and any urban flooding that would have occurred if the city was hit with what Houston had experienced—a one-thousand-year storm-and-flood event.[43]

Suzanne Scott, general manager of SARA, drew an important conclusion from this incomplete data. Noting that it was clear that San Antonio could not build its way out of flooding, which was a striking admission for any river authority to make, at least the agency and its constituents could take advantage of "the tools—like the risk map—to help people understand and mitigate risk."[44] Embedded in her cautionary words is an acceptance of the community's enduring vulnerability, its precariousness. Her admonition, much like the floodpool data that SARA workers had inscribed on the Olmos Dam gatehouse, acts as did their handwriting on the wall: it made legible the reality that San Antonio occupies a hybridized watershed, at once built and natural, controlled, fluid, and dangerous.

NOTES

1. "Olmos Detention Dam Linking San Antonio and Alamo Heights Formally Dedicated by Officials," *San Antonio Express*, August 28, 1926, 26; "Tobin Opens Olmos Dam Drive," *San Antonio Light*, December 11, 1926, 1, 2; "Scenes at Dedication of $1,600,000 Olmos Dam," *San Antonio Express*, December 12, 1926, 26.
2. The engineers' reports in "Letters and Photographs Pertaining to Construction of Olmos Dam, Texas," in Special Collections and University Archives, WRCA G248 E7, University of California, Riverside.

3. *Olmos Dam Gatehouse: Writing on the Wall* (San Antonio: San Antonio River Authority, December 17, 1980).
4. White, *The Organic Machine*, 112.
5. Way, "Introduction," 3.
6. Miller, *Westside Rising*, 4–6; Petersen, "San Antonio," 17–40.
7. Bartlett, "The Flood of September 1921," 355.
8. Ellsworth, "The Floods in Central Texas," 5.
9. Burnett, *Flash Floods in Texas*, 28–49.
10. Miller, *San Antonio*, 3–11.
11. Major floods wracked the city in 1819, 1845, 1865, 1869, 1880, 1885, and 1889; then in 1894, 1899, and 1900; another run occurred in 1903, 1905, and 1907; and then two in 1913, another each in 1914 and 1915, before the megaflood of 1921.
12. A quarter league is about one mile; Ramón quoted in Stothert, *The Archaeology and Early History*, 54.
13. Olmsted, *A Journey through Texas*, 149.
14. Prescott, "San Antonio de Bexar," 835.
15. *Hanford's Register for 1878* quoted in Everett, "San Antonio Welcomes the 'Sunset'—1877," 56.
16. *San Antonio Daily Express* quoted in Zunker, *A Dream Come True*, 79.
17. Sibley, *George W. Brackenridge*, 141.
18. "Start the Big Idea," *San Antonio Express*, February 5, 1911, 1.
19. Fisher, *Saving San Antonio*, 192–220.
20. Quoted in Fisher, *American Venice*, 35.
21. Fisher, *American Venice*, 92–93.
22. Hugman quoted in Fisher, *American Venice*, 93, 95.
23. Johnson, *In the Loop*, 247–49.
24. Madla, "Landscape Narratives," 253–54; Johnson, *In the Loop*, 304–15; Miller, *San Antonio*, 113–14, 245–48.
25. Yviand Serbones, "Economic Impact Studies Reinforce Value of San Antonio River Walk Extension," *River Reach*, Fall 2014, 5, https://www.sariverauthority.org/sites/default/files/2019-06/newsletter_fall2014.pdf.
26. Char Miller, "These Spurs Are Made for Walking," *TerraFirma*, June 23, 2014, https://terrafirmastory.com/2014/06/23/these-spurs-were-made-for-walking/.
27. Fisher, *American Venice*, 177–99.
28. Madla, "Landscape Narratives," 256.
29. Miller, *Westside Rising*, 150–82.
30. Colin McDonald, "First Steps on Bike Trails," *San Antonio Express-News*, June 11, 2008, 6A; Vianna Davila, "Ground Broken for Greenbelt," *San Antonio Express-News*, December 17, 2009, 1.

31. Vianna Davila, "Residents Cheer First Linear Park on West Side," *San Antonio Express-News*, November 20, 2010.
32. Larry Clark, landscape architect and former chair of the Open Space Advisory Board, in discussion with the author, December 10, 2020; Jeremy T. Gerlach, "Construction Starts on Martinez Creek Hike/Bike Trail," *San Antonio Express-News*, April 29, 2015, 65; Scott Huddleston, "Voters Approve Aquifer, Linear Parks Propositions," *San Antonio Express-News*, May 10, 2015, 17; Colin McDonald, "Hike-and-Bike Trails Waiting for Expansions," *San Antonio Express-News*, November 26, 2010, 1, 6; Davila, "Residents Cheer First Linear Park on West Side."
33. John Welch, "Peak Sees Linear Parks as Part of His Legacy," *San Antonio Express-News*, November 3, 2000, 59; Lainey Berkus, "He's on the Move with Tools to Keep Going," *San Antonio Express-News*, July 24, 2014, 32; Ray Knox, "Funding Needed to Maintain Parks," *San Antonio Express-News*, August 5, 2013, 11.
34. "About the Creeks," Westside Creeks Restoration Project, October 19, 2023, https://www.westsidecreeks.com/about-the-creeks/; *Westside Creeks Restoration Project Conceptual Plan*, San Antonio River Authority, 2011, accessed October 19, 2023, https://www.edwardsaquifer.net/pdf/Westside_Creeks_Restoration.pdf.
35. *Westside Creeks—Final Draft Report*, San Antonio River Authority, February 18, 2020, https://www.westsidecreeks.com/wp-content/uploads/2020/03/WCROC_02182020_WSC_ALL-PRESENTATIONS.pdf, 12–16; *Westside Creeks Restoration Project Conceptual Plan*, 2011.
36. Hamilton and McDonald, "Decision on New River Trail Approaches," *San Antonio Express-News*, December 15, 2008, 1, 8.
37. "Confluence Park," San Antonio River Foundation, accessed October 19, 2023, https://sariverfound.org/confluence-park/.
38. "Confluence Park," Lake/Flato Architects, accessed October 19, 2023, https://www.lakeflato.com/eco-conservation/confluence-park/?project=open.
39. *South Texas Floods, October 17–22, 1998* (Washington DC: National Oceanic and Atmospheric Administration, February 1999), 1–15; *Floods in the Guadalupe and San Antonio River, Texas* (Restin VA: United States Geological Survey, October 1998); Sig Christenson and Adolfo Pesquera, "3 Killed as Floods Hit S. A. Area," *San Antonio Express-News*, October 18, 1998, 1A; Adolfo Pesquera, "Bad Weather Forces Residents, Rescuers to Evacuate," *San Antonio Express-News*, October 18, 1998, 16A.
40. *City of San Antonio Flood Mitigation Plan*, City of San Antonio Office of Emergency Management, December 2000, https://www.saoemprepare.com/Portals/16/Files/Plans/planHMAP.pdf?ver=2017-03-01-001439-567, appendices K and W; Iris Dimmick, "Flood Models Show that Olmos Dam Would Overflow in Harvey-Like Storm," *San Antonio Report*, November 1, 2017.

41. Furl et al., "Hydrometeorology," 90–104; Kelsey Bradshaw, "UTSA Study Finds Urbanization Made 2015 Memorial Day Weekend Flood Worse," *San Antonio Express-News*, February 5, 2018, 1.
42. Zhang et al., "Urbanization," 384–88.
43. Dimmick, "Flood Models."
44. Dimmick, "Flood Models"; "SARA Risk Map," San Antonio River Authority, accessed October 19, 2023, https://sara-tx.maps.arcgis.com/apps/webappviewer/index.html?id=0b13614f13124257bfe589a459ba84fe.

BIBLIOGRAPHY

Bartlett, C. Terrell. "The Flood of September 1921, at San Antonio, Texas." *Transactions of the American Society of Civil Engineers* 85, no. 1 (January 1922): 354–65.

Burnett, Jonathan. *Flash Floods in Texas*. College Station: Texas A&M University Press, 2009.

Ellsworth, C. E. *The Floods in Central Texas in September 1921*. Washington DC: U.S. Geological Survey, 1923.

Everett, Donald E. "San Antonio Welcomes the 'Sunset'—1877." *Southwestern Historical Quarterly* 65, no. 1 (July 1961): 47–60.

Fisher, Lewis F. *American Venice: The Epic Story of San Antonio's River*. San Antonio TX: Maverick Books/Trinity University Press, 2015.

———. *Saving San Antonio: The Precarious Preservation of a Heritage*. Lubbock: Texas Tech University Press, 1996.

Furl, Chad, Hatim Sharif, Jon W. Zeitler, Almoutaz El Hassan, and John Joseph. "Hydrometeorology of the Catastrophic Blanco River Flood in South Texas, May 2015." *Journal of Hydrology: Regional Studies* 15 (February 2018): 90–104.

Guerra, Claudia R., ed. *300 Years of San Antonio and Bexar County*. San Antonio TX: Trinity University Press, 2018.

Johnson, David R. *In the Loop: A Political and Social History of San Antonio*. San Antonio TX: Maverick Books/Trinity University Press, 2020.

Madla, David. "Landscape Narratives and the San Antonio River." In *River Cities, City Rivers*, edited by Thaïsa Way, 241–66. Cambridge MA: Harvard University Press, 2018.

Miller, Char. *San Antonio: A Tricentennial History*. Austin: Texas State Historical Association Press, 2018.

———. *Westside Rising: How San Antonio's 1921 Flood Destroyed a City and Sparked a Latino Environmental Justice Movement*. San Antonio TX: Maverick Books/Trinity University Press, 2021.

Olmsted, Frederick Law. *A Journey through Texas; Or, a Saddletrip on the Southwestern Frontier*. New York: Dix, Edwards, & Co., 1857.

Petersen, James F. "San Antonio: An Environmental Crossroads on the Texas Spring Line." In *On the Border: An Environmental History of San Antonio*, edited by Char Miller, 17–37. San Antonio TX: Trinity University Press, 2005.

Prescott, Harriet Elizabeth. "San Antonio de Bexar." *Harper's Magazine* 55 (June–November 1877): 831–50.

Sibley, Marilyn Macadams. *George W. Brackenridge: Maverick Philanthropist*. Austin: University of Texas Press, 1973.

Stothert, Karen E. *The Archaeology and Early History of the Head of the San Antonio River*. San Antonio TX: Incarnate Word College, 1989.

Way, Thaïsa. "Introduction." In *River Cities, City Rivers*, edited by Thaïsa Way, 1–12. Cambridge MA: Harvard University Press, 2018.

White, Richard. *The Organic Machine: The Remaking of the Columbia River*. New York: Hill and Wang, 1995.

Zhang, W., G. Villarini, G. A. Vecchi, and James A. Smith. "Urbanization Exacerbated the Rainfall and Flooding Caused by Hurricane Harvey in Houston." *Nature* 563 (2018): 384–88.

Zunker, Vernon G. *A Dream Come True: Robert Hugman and San Antonio's River Walk*. San Antonio TX: GraphicShop, 1994.

5 From Floodplain to Kayak Park

Recreation, Restoration, and Economic Development along the Boise River

Jennifer Stevens

This morning my alarm woke me at 6:00 a.m. I laced up my shoes and headed to the river. My run started with three miles upstream on the north side of Boise River Greenbelt to get the harder part out of the way and finished with three miles downstream on the south side, enjoying the river's gentle grade on my return. Even at that early hour and in the springtime desert chill, kayakers and surfers lined up in the water waiting for their turn at the wave at Boise's Whitewater Park, while observers stood at the overlook. I was reminded, as I often am, that this was not the river of the 1970s, and that urban leaders' efforts merged fortuitously with a developer's missteps to clear the way for a beautiful river to grace an urban space.

Since the early 2000s, Boise, Idaho's capital and largest city, has been sitting atop a growing number of "best of" lists.[1] In 2019 livability.com ranked Boise the best place to live in the entire country. Similarly, in 2020 the housing website curbed.com listed it number two on its list of best cities to live in the United States. It has also been ranked the eighth most physically active "large" city, the best place for millennials, and the best place to buy a home. The list goes on. Without a doubt, folks who have called Boise home for any meaningful amount of time are a little stunned by this moment that Boise is experiencing. Of course, they love their city, but why now? What explains the sudden national interest in a place that has been an isolated, agriculturally focused outpost for nearly its entire modern existence? A deeper dive into the lists' descriptions of the city provides a glimmer of an answer. For example, curbed.com simply described Boise as "a mountain vista that offers the outdoors and new business opportunities."[2] Who wouldn't agree that it sounds nice? But Boise's modern reputation as an outdoor mecca is new. It was cultivated, and it was curated—like a museum exhibit—born from the confluence of belated comprehensive planning, the tragedy of Boise's approach to urban renewal, and the modern environmental movement. And now, after fifty years, Boise is reaping the benefits.

The Boise River runs right through the middle of the city limits, just a short walk from the downtown core. It emerges from the mountains that flank the city's northeast boundary and flows more than fifty miles to the west where it meets the Snake River on the valley's western edge. Today's residents, most of whom never experienced the river's annual natural rhythms, take their recreational playground for granted and consider stories of the Boise River's historic springtime flooding to be almost mythical. Black-and-white flood photos that periodically appear on Boise history-related social media pages never fail to garner digital oohing and aahing. After all, it's hard for Boise's modern citizen—who likely migrated recently from another state—to imagine a time when the river didn't showcase the city's contemporary green identity, an era when locals considered the river to be nothing more than an out-of-control menace in desperate need of taming, or a dumping zone to carry away the filth discharged by progress.[3]

In truth, today's favorite kayaking spot was yesterday's busiest slaughterhouse, steel fabrication plant, or metal shop, and most Boiseans don't understand just what it took to enjoy what they have now. Boise barely survived its own catastrophic urban renewal story, emerging from it bereft of identity, if it had ever had one. But there was a glimmer of hope that might revive the city, for while Boise was tearing itself down, the environmental movement was simultaneously flourishing, providing an opportune time for a city searching for itself to connect anew with its natural amenities. Unlike many Rust Belt cities, whose path forward was constrained by a history associated with industry, Boise's absence of a similar past permitted its quest for a new identity to become a source of resilience in the face of jarring changes. The Boise River's past as a working river was fading into obscurity, but the stream formed the crux of a new identity, envisioned by a few forward-looking individuals as the centerpiece of a refurbished western city.

Balancing economic growth, river recreation, and resource conservation, however, turned out to be a complex task. As with all best-laid plans, unintended consequences associated with making a recreation-rich river—manipulating flows and the floodplain to encourage human proximity to the water—complicated the outcome, since the "naturalness" of the river's ecology was compromised by urban dwellers' desire to be near it.[4] While the necessity of regulating the Boise River created economic change in the valley, it also became the unwitting source of resilience in a city searching for an identity.

The River, Take One

The history of the Boise River following European arrival is not so different from other urban rivers. Before the European fur traders arrived, the Shoshone Tribe used the Boise Valley seasonally. Fur trading explorers appeared at the turn of the nineteenth century, while permanent European (and other) settlement began around the 1860s. The river's unpredictability during all that history was, in fact, fairly predictable. Settlers in this semi-arid landscape recognized the dangers of settling too close to the river for fear they would be washed away in a spring freshet, yet they depended on its water and needed access to it. In the years before hydrologists conducted snow studies and irrigation engineers built dams to hold back the runoff from annual spring warming, residents had no choice but to accept that the river carried more than just water: it also carried risk.

After the city's founding in 1864, settlers gradually populated land along the waterway, and a diverse economy followed. Extractive industries like timber and mining drove regional growth, and Boise became one of the hubs through which raw materials from such industries passed. As a result, businesses that supported those industries evolved into the foundation of Boise's economy for nearly a century. However, the historic floods that came each spring in the first half century of settlement were destructive and inconvenient. Furthermore, the climate's contrasting extreme of late summer dry spells made farming a challenge. The extremes of too much and too little water prompted the newly created Bureau of Reclamation to target the Boise River for one of its earliest projects. The agency completed its first storage dam, Arrowrock, on the South Fork of the Boise River in 1915, which remained the highest dam in the world until 1932. However, when the Boise River overtopped Arrowrock in 1943 and caused massive flooding throughout the city, the U.S. Army Corps of Engineers and the U.S. Bureau of Reclamation teamed up to build a second, larger capacity reservoir just downstream from Arrowrock on the river's main stem.[5] Construction on Lucky Peak Dam concluded in 1955, creating an intensely manipulated river, one whose ebbs and flows only barely resembled those of a century earlier.

Over a mere hundred years, humans dramatically altered the Boise River with dams that both stored spring runoff and reduced flood risk. Controlling the river's flow allowed engineers to narrow and straighten the channel. During these same years, farmers and ranchers expanded and shipped food to mining and timber towns, and entrepreneurs erected sawmills, dairies, and slaughterhouses in and around Boise. Additionally, the town's geographic isolation from other urban centers prompted the birth of a small industrial sector, comprised first of quarries and concrete plants, and then of large machine shops, iron foundries, and steel fabrication plants, the products from which served the larger regional economy. The controlled river channel facilitated the siting of these new industries on or very near the banks of the Boise River.

Throughout the twentieth century, these changes allowed people and businesses to treat the Boise River as the valley's own sink into which they poured every waste imaginable. From the mining waste that flowed out of towns situated along the river's tributaries, to the sawmill at the valley's upstream end, to the slaughterhouses and fabricating plants dotting its course, and to the pesticide runoff from farms near its mouth, the Boise River received the toxic discharge of "progress" for decades before the government regulated such things. This was no Cuyahoga River, but neither was it pristine.[6]

While none of the story thus far is terribly unique, the more contemporary history of the Boise River is less typical. Boise was still a very small and faceless place when Congress passed the 1972 Clean Water Act, which aimed in part to clean up all rivers in the United States. Part of the cleanup included rules prohibiting industries from indiscriminately discharging their waste into the nearest waterway. In the face of these regulations, most cities—especially those in the Rust Belt—battled to maintain the industrial jobs they stood to lose rather than reimagining a new role and identity for

4. Boise Industries, Gate City Steel, originally Olson Manufacturing. Courtesy of Idaho State Archives, photo 78-37.183.

the cleaned-up rivers. Those cities were deeply dependent on leaders' past decisions, which had created urban identities and infrastructure around hard industries such as steel and paper manufacturing. Consequently, a great number of urban rivers became industrial graveyards, as their cities simply abandoned them; efforts to restore or reenvision these waterways came much later, a final gasp only after the industrial loss was complete and permanent.[7] At least some leaders in western urban centers reacted similarly to their counterparts in the Rust Belt, despite the fertile ground for river restoration and reimagination they may have found in the West's long reputation as a health and recreation magnet. They, too, failed to immediately recognize or embrace the potential these waterways offered to provide a lifeline of resilience in the transition wrought by deindustrialization and globalization. Los Angeles, for example, restored only small sections of its long namesake river, leaving most of it encased in concrete even today, well into the twenty-first century.[8] Seattle, too, invested in keeping the Duwamish River—which drains into Elliot Bay—industrial. But Boise leaders—unconstrained by a clearly defined path or identity—were different and embraced a starring role for the Boise River in the city's identity early in the environmental era. The constrained channel had created a river flowing neatly between two defined banks instead of across a wide floodplain and had historically facilitated the siting of industry along its course. But in this new era, the channelization of the river permitted accessible recreation along its banks. What became the Boise River Greenbelt vision—and later, the River Plan—brought people back to the river, changing citizens' relationship to it and enhancing their appreciation of the resource. The complexities of accommodating recreation, river ecology, and economic growth inevitably resulted in compromise along the river, but the river's role in Boise's modern identity would be assured.

The Boise River and Boise's Version of Urban Renewal

The city's elected and business leaders were hatching plans to capitalize on Boise's natural amenities as early as the mid-1960s, as they searched for a way to revitalize the downtown. Like other cities in the postwar period, Boise's citizens were fleeing the urban core to resettle in the surrounding foothills and suburban western valley.[9] These demographic trends led to increasing concern over the downtown's crumbling buildings and lackluster business district. The private sector, hoping to stem population loss and slow decentralization, formed the Boise Planning and Development Committee in late 1964 with the tagline, "A non-profit corporation to create a dynamic plan and sound guide for greater Boise development."[10] The city government of Boise mimicked these private urban renewal efforts by creating the first planning department at city hall and chartering the Boise Redevelopment Agency, or BRA, in 1965.

That year, the private sector Planning and Development Committee invited the Urban Land Institute to host a panel to discuss development in Boise. The panel emerged with several recommendations that appear disparate in retrospect: on the one hand, the group recommended more surface parking and the creation of superblocks in the downtown core. On the other hand, they urged the establishment of a footpath along the river, which they called the Greenbelt, to "further enhance Boise's excellent park system."[11] At first, no one voiced what now seems patently obvious: these recommendations seemed at odds with one another. Destroying historic buildings (which are unique and bestow character) to provide surface parking (which is a faceless commodity), while at the same time trying to create urban character through a new riverside footpath and a nod to recreation, were contradictory. The planners and business owners viewed Boise's old downtown buildings as decayed and their blocks as blighted; they cheered the dismantling of these buildings that comprised the city's core. As a result, Boise simultaneously allowed its downtown to be ripped apart while building up its river park. Unexpectedly, the deconstruction of the city became an opportunity for development, marketing of the city's natural resources, and the construction of a new urban identity. And so, while Boise's urban renewal efforts were similar in many ways to its sister cities throughout the country—demolish the old and replace with the new—Boise's efforts included something many other such plans had not: a commitment to the development of its natural amenities.[12]

As American cities go, Boise was slow to adopt comprehensive planning. It passed its first plan in the 1960s, and consequently, the language and recommendations reflected Americans' changing environmental perspectives. Nearly four decades after the U.S. Supreme Court upheld the constitutionality of municipal zoning codes and (implicitly) comprehensive planning in 1926, Boise finally contracted with California planning firm Atkinson and Associates to write the capital city's first comprehensive plan in 1963.[13] On the whole, Boise's new plan was nothing out of the ordinary. It defined the city's areas of business and industry, noted its topography and natural features, and generally provided suggestions for guiding future development.

One sentence in Boise's Metro Plan, however—buried in the middle—had a disproportionate impact on the city's future: "Acquire land along the Boise River so as to create a continuous green belt of public lands stretching along the river throughout the entire length of the community."[14] That plan also brought to the city government's attention what had previously been overlooked: "Boise City has been favored by natural, scenic surroundings of rivers, mountains and dramatic civic beauty forming an essential part of the Boise City heritage."[15] Soon after adopting the plan in 1964, the city engaged a local firm called Planning Research West, which completed the city's first Boise River Greenbelt Comprehensive Plan and Design in

1969. The city council officially adopted it that year.[16] The Greenbelt plan opened by praising the city for having "chosen to aggressively attack the implementation program of the Boise River Greenbelt" and cited exponential population growth throughout the United States as a reason why the "agonizing reappraisal of the highest and best use of prime natural resources has seen the pendulum swing to preservation and reconstitution of the original resource." The city's 1967 application to obtain federal matching funds from the Land and Water Conservation Fund to build the Greenbelt noted that "time . . . is about to run out and unless a positive program of acquisition is undertaken immediately a priceless opportunity may be lost forever."[17] Similar reckonings were occurring on a smaller scale in other places and in other cities, but none that would have such an impact on a city's wholesale identity.[18]

Support for Greenbelt implementation appeared to be widespread. The plan called for the city to acquire three thousand feet of riverbank land stretching for ten river miles along the north side of the Boise River. Although its precise cost remained unclear, the price for land acquisition (or rights of way or easements) was expected to be astronomical. Nevertheless, enthusiasm for the riverfront amenity became a litmus test for candidates running for election to city office in fall 1969. Eighteen candidates threw their hats in the ring for three city council seats, and of all the issues that could have been probed in that year of chaotic national issues, the local newspaper asked the candidates only three questions, one of which was, "Do you consider the greenbelt important to improving the quality of our Boise environment?"[19] All three candidates who succeeded at the ballot box on November 4 answered a resounding, "Yes."[20]

Boise's Economic Transition

Over the next few years, the city implemented the Greenbelt through private property purchases and rights-of-way negotiations on lands adjacent to the river. While not overtly recognizing that the Greenbelt project was part of Boise's urban renewal story, policymakers at the city nevertheless observed that starting the project from the downtown core and building up and downstream from there was the best way to get people to come back to the city. In places where adjacent landowners expressed unwillingness to sell, the city used the power of eminent domain to acquire the parcels, or rights-of-way. By 1975 when the first paved section opened just south of downtown, $500,000 had been spent on acquisitions.[21] The following year the city enlisted Ada County to contribute funding and help facilitate easements on lands that lay outside the city and in the county's jurisdiction. By then, the demand for this new amenity had become increasingly loud and was coming from a new segment of the city's population and economy.

Boise's transition to an information/service economy is an important part of this story. It began with California-based Hewlett Packard's (HP) 1973 decision to open its first printer division in Boise. After looking for a place that offered affordable housing, diversity of religion and ethnicity, low crime, and high quality of life, HP's site-selection committee chose Boise over Spokane, Washington, and Corvallis, Oregon.[22] HP's Ray Smelek, who led the move, said he felt an affinity for the place immediately, and the availability of golf courses and cheap ski passes sealed the deal.[23] Soon after HP's arrival in Boise, four young men breathing the fumes of laughing gas in the basement of a dentist's office founded Micron Corporation, using investment money provided by some of Idaho's best-known entrepreneurs to manufacture memory chips for computers.[24] Micron, founded in 1977, went on to become a multi-billion-dollar, multinational corporation. Its success—along with HP's—transformed Boise into a mini-Silicon Valley, minus the traffic hassles and expense. Boise also offered accessible amenities that made the valley an attractive place to raise a family. Boise's evolving identity would be shaped by both the people moving there to work in the information economy and the city marketing its natural amenities. In fact, Boise's new "green" identity was aimed at a very specific profile—a white, middle-class professional who would be attracted to new jobs in the valley because of the "quality of life" offered by outdoor amenities like river and mountain recreation.

The positive feedback generated by implementation of the Greenbelt plan was not enough, however, to obscure negative feedback generated by the destruction caused by urban renewal. It was particularly embarrassing when, in November 1974, *Harper's Magazine* ran an article by L. J. Davis, a Boise native who had returned for a visit from his home on the East Coast and was shocked by what he saw. In his locally infamous article, Davis observed that "Boise stands an excellent chance of becoming the first American city to have deliberately eradicated itself."[25] Davis noted that although downtown clearance—a common urban renewal effort—had finally become "unfashionable . . . because it has never worked," the process had nonetheless continued apace in Boise. He followed with, "If the only thing you can do when you come downtown is park your car where the place *used* to be, most people are pretty much inclined to say the hell with it" and head to the nearest suburban shopping mall (emphasis added). Turning downtown Boise into what Davis called "a prairie of parking lots" only destined the downtown to suffer even more fully the forces that led people away from downtown to begin with. Finally, Davis asserted that the city of Boise was "afflicted with an inferiority complex."[26]

It was true. Elected leadership was paralyzed by the apparent trade-offs between modernization, urban character, and identity. Whatever had comprised Boise's past character and identity was gone or in the process of disappearing, but the public still wasn't quite sure what the new identity would be. Information workers would soon be coming in droves with the growth of HP and Micron, but those same years would be marked by the departure of the machine shops, slaughterhouses, and a steel fabrication plant, businesses that were often forced to either shut down or depart due to the expense of waste disposal or the city's new comprehensive plan that discouraged industry from locating in or around the city center. As the city stumbled toward remaking itself, it trod a thin line between emphasizing natural-amenity recreation and ignoring natural-resource protection, all the while tearing down its core. Although Davis's observations were a mere snapshot during a time of transition, Boise's experiment with building a green identity—which started with the Greenbelt and coincided with destruction of its past—was not yet assured. The next forty years would bring new twists and turns to the story.

The River, Take Two

Boise succeeded in creating a new identity beyond what anyone could have predicted. The "best of" lists are one indicator, population growth the other. Without fail, the articles or blurbs that have described Boise on these lists identified the city's quality of life, always mentioning the now 22-mile Greenbelt that flanks both sides of the river, as well as 190 miles of a city trail system that snakes throughout the foothills, giving residents many convenient opportunities to escape city life. Downtown eventually bounced back, too, with a thriving retail and restaurant sector and a gradual eradication of parking lots. Unsurprisingly, being on these lists attracted more people to Boise, which forced Boise leaders to think anew about the river itself, not just the role that the river played in the city's identity. In other words, drawing people to the resource was revealing new problems they hadn't originally considered: how could the city protect the river while simultaneously encouraging people to use it?

Boise's population boom started in the 1970s and never let up. After more than twenty years of stagnant population growth, the rise of the technology industry and Boise's new amenity-driven focus led to rising in-migration during the decade between 1970 and 1980. The upward trajectory continued even despite the early 1980s recession, and the city continued clocking double-digit percentage gains every decade (36 percent from 1970 to 1980, 23 percent from 1980 to 1990, 41 percent from 1990 to 2000, and 16 percent from 2000 to 2010).[27] Then, between 2000 and 2020, the population of Boise grew from 176,802 to 229,993, a 20 percent increase. More astounding yet are the numbers from the county in which Boise is situated. During those same twenty years, Ada County grew by a whopping 64 percent, from 300,904 residents to 494,967, creating greater recreational demands on the river.[28] The Boise River Greenbelt expanded slowly but steadily during this time, driving (and driven by) an intensified human desire to be near the water. Manipulation of river flows through the construction of upstream dams between 1910 and 1950 allowed the Greenbelt to exist, and by 1990 it stretched for twenty-one miles. The last section was finally completed in 2016, when the parks department dug the final tunnel under Main Street, marking the last and final connection.[29]

It is unsurprising that as the Greenbelt became a reality, the river became more than just a recreational destination for daily use. Businesses, for one, recognized the value of situating their offices with windows overlooking the newly cleaned-up banks. Office buildings sprang up along the river in the 1980s, housing law firms, dentists' offices, and real estate companies.[30] Boise-based supermarket chain Albertson's constructed a new headquarters building along the south bank in 1980 (aptly named River View Plaza), as did other large corporations like Ore-Ida Foods and H. J. Heinz.[31] Employees of these businesses could head outside during their breaks and stroll along the riverfront or eat their lunches at the strategically placed benches and tables. Restaurants exploited citizens' desire to eat adjacent to the river, as well. The developments gradually began to bring workers back to the city, arguably one of the initial justifications for the pathway.[32]

Some residential home developers also took their chances building subdivisions along the river close to downtown, banking on the continuing influx of information workers who wanted to live both close to natural amenities and downtown's remaining attractions. One of the most significant of these was proposed by Boise developer Peter O'Neill. He envisioned River Run as a 650-unit development on 120 acres along the south bank of the river. The land was formerly used as a cattle pasture and located just three miles east of the state capitol, which could be reached by a leisurely stroll or commute-speed bike ride along the new Greenbelt.[33] A 6-acre feature called Heron Lake remained on the property, named for the avian species that lived in several rookeries along this urban stretch of the Boise River. A series of waterways leading to the river also remained, most significant of which was a half-mile-long flood-relief channel that paralleled the river. O'Neill had a unique vision for the development. While fellow developers were simultaneously proposing subdivisions in other fragile environments in the Boise foothills that required flattening mountains, O'Neill saw an opportunity to help restore the land and create a more ecologically friendly setting while also capitalizing on the location close to downtown.[34] He hoped to keep the waterways flowing throughout the summer, and as an avid fisherman, to see the subdivision's streams running with trout. To make his vision a reality, O'Neill hired consultants, won the support of Idaho Fish and Game, and eventually installed fish ladders and a meandering set of waterways throughout the subdivision, which included pools, riffles, and spawning beds. He also deepened Heron Lake and increased the length of the waterways through the subdivision.[35] The River Run development was constructed in four phases between 1981 and 1989, and the housing there retains its values today, remaining a highly desirable location to live.

Not all riverside developments treated the natural environment with equal care, however, and as more people and development were drawn to the riverbank, there was a growing sense that protection might be needed. Townhouses, apartments, office buildings, single-family homes, and retail developments were going up fast, as close to the river and Greenbelt as they could get. As new and long-standing residents began to reverse the decentralization trend and move back toward the city center and the Greenbelt grew in popularity, some leaders at city hall and citizens in the community began to recognize the need to not only draw people to the river, but to also create a system to protect its fragile ecosystem. As such, the city commissioned the Boise River Fish and Wildlife Habitat Study in 1983 to provide a check on the growing interest in riverside development. The study offered a classification of lands according to their wildlife and natural value, leading to a larger planning effort and to the city's adoption of the Boise River Plan in 1985, followed by an implementing ordinance in 1993.[36]

By dividing lands into well-defined categories, the Boise River Plan provided city planners and potential developers with a common language and set of legal restrictions depending on the land's classification. Class A lands were those most worthy of preservation practices, providing "extremely important habitats for fish and wildlife and for flood control and protection."[37] The code protected class A lands by restricting uses for several hundred feet around them. The lands and waters that fell under this classification included floodways, diverse plant community areas, heron rookeries and bald eagle wintering habitat, islands in the river, wetlands, and riparian plant communities. The plan described class B lands as being "moderately important" and included agricultural lands, gravel pit ponds and small lakes, and intermittent tributary streams. Finally, class C lands provided the least opportunity for preservation or enhancement because of incompatible uses. Regulations were most restrictive for class A lands, of course, and included a minimum 70-foot setback for any structures or parking areas. Heron rookeries required a 300-foot buffer, while eagle habitat demanded a 200-foot setback from the high-water mark, set at 6,500 cubic feet per second.[38] Passage of the Boise River Plan in 1985, and especially the 1993 ordinance that implemented the plan, marked important turning points for Boise by erecting burdensome regulations that made construction in the floodplain extremely difficult by design. The Greenbelt might draw people in, but the Boise River Plan and ordinance would make sure people's footprint was light and transient. Eventually.

Implementing the Plan

My runs along the Boise River often take me past a new park—completed in 2021—just downstream of the Fairview Bridge and two miles west of

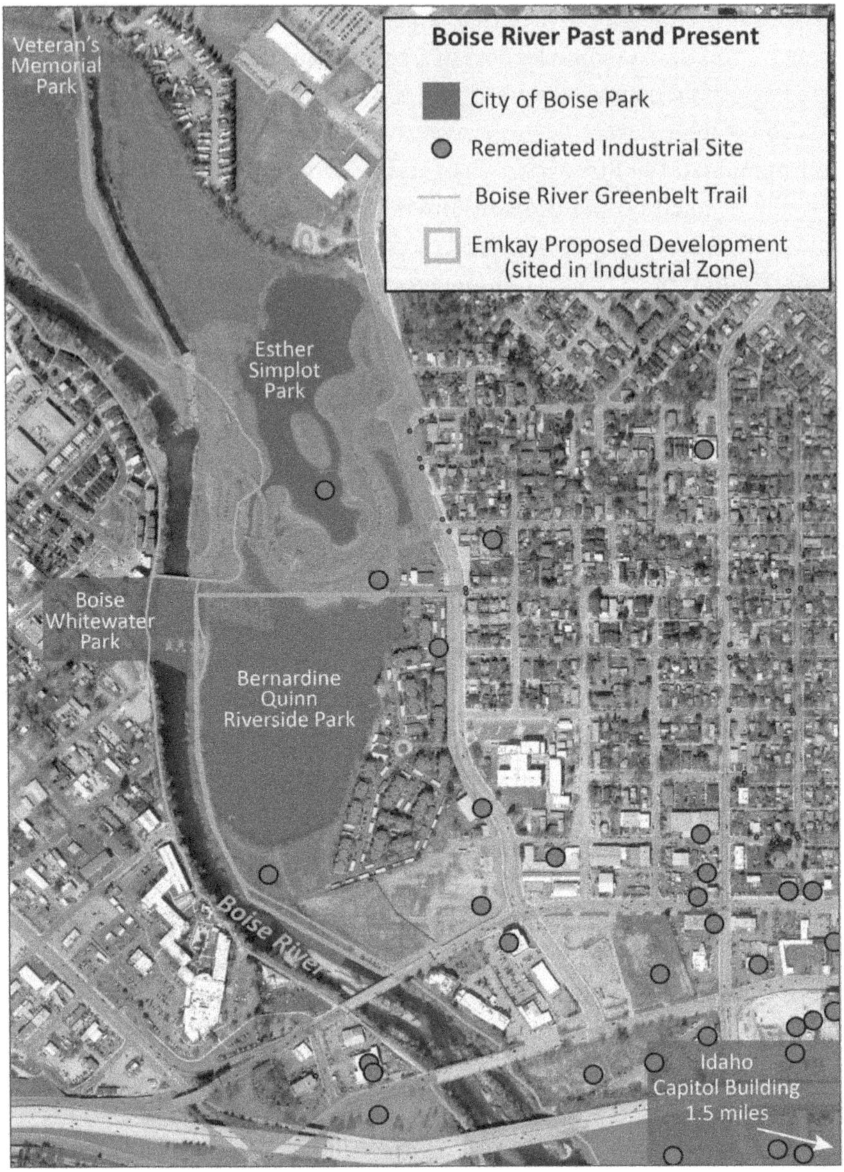

5. Map of the Boise River without the mall. Created by Amalia Baldwin, Stevens Historical Research Associates.

the state capitol. Bernardine Quinn Riverside Park lies adjacent to the north riverbank and to the Esther Simplot Park just downstream, both on what used to be industrial lands.[39] To the north, a vast parking lot—the final remnant of a former car dealership—is a reminder of the area's past. The Greenbelt connects them all, providing access to the parks' large recreational ponds, which are fed by the Boise River. Today, Boise's new in-river Whitewater Park adjoins these land-based parks. The Whitewater Park was completed in recent years when an irrigation structure needed replacement and the city worked to accomplish two goals at once. The whole area teems with people from dawn until dusk, while Canada geese, blue heron, wood ducks, mallards, mergansers, osprey, sandpipers, bald eagles, red-wing blackbirds, gulls, mule deer, beavers, and river otters call the area home. It almost wasn't like this.

This two-mile stretch of river housed industrial operations that dumped their waste into the stream for many years. Today's ponds were yesterday's gravel pits, and despite the city's new River Plan, the floodway's industrial legacy—still visible on the landscape—made it difficult in 1985 to see the value of even class A lands and easy, instead, to envision new construction or development. Urban renewal's downtown destruction had left Boise with no strong retail center, and in the resulting struggle to attract people back to the urban core, the city faced a decision about where to build what would become the metropolitan area's shopping destination. The difficulties of balancing river recreation, river preservation, and economic development came into sharp focus when, in late 1985, one of three would-be mall developers proposed a project at this site on the river's edge. Boise-based Emkay Development Corporation, Inc. (a spin-off from construction company Morrison-Knudsen) proposed to replace these former gravel pits, slaughterhouses, and concrete companies with a regional shopping mall snug up against the Boise River, immediately adjacent to the Greenbelt.

Emkay's application to rezone the parcel from its existing industrial designation to a commercial zone came just months after Boise adopted the River Plan and provided a testing ground for the city. The developer's plan, submitted in October 1985, was to construct a seven-hundred-thousand square-foot shopping mall.[40] Despite the proposal's apparent compliance with the Boise Metro Plan's requirement that a regional shopping center be "surrounded by urban development and located a short distance from

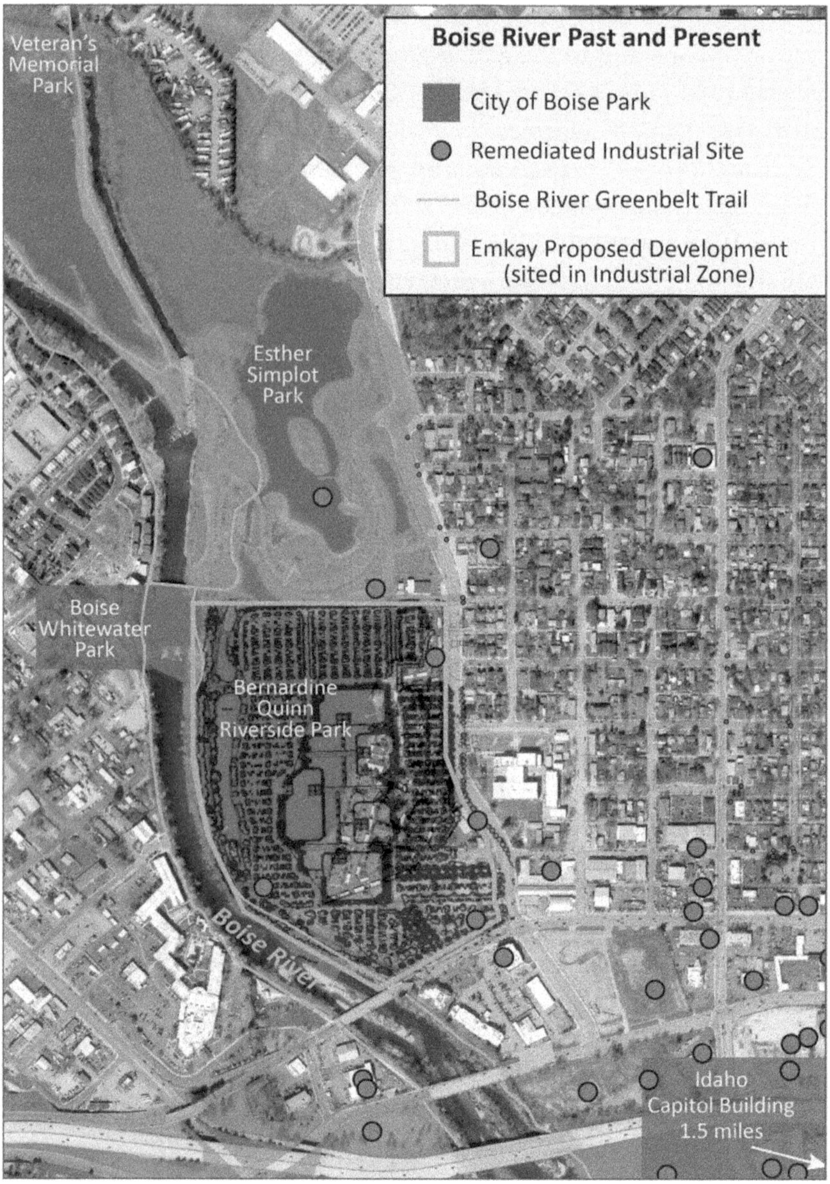

6. Boise River with Emkay Mall proposal overlaid. Created by Amalia Baldwin, Stevens Historical Research Associates.

the central business district," the site's location along the river constrained the proposal from the start.[41] Its first challenge was that a large portion of the subject parcel was situated in the designated floodway of the Federal Emergency Management Agency (FEMA), which restricted construction options and rendered more than half of the site unbuildable.[42] Furthermore, Emkay would have to comply with the new Boise River Plan, which underscored the federal regulations by locally prohibiting any construction in the class A floodplain lands except when required for public necessity, including emergency access, recreational use, wildlife habitat, and gravel extraction.[43] The developer included a figure, taken from the government's 1868 survey, which clearly depicted the river's original location on this precise spot.[44] Not surprisingly, class A lands covered a third of the site as well. Despite the law and the River Plan's clear language, Emkay's proposal pushed all of these boundaries by proposing buildings and asphalt parking lots in both the floodway and the floodplain, and the filling of half the site's class A lands.

This application was a flashpoint in Boise's then twenty-year-old effort to place the river at the center of the city's identity; but there was a distinct clash of two important policies. On the one hand, the city was desperate for citizens to return to the urban core and believed that a shopping center could accomplish this goal. On the other hand, the city had committed to river resource protection by passing the River Plan. It was not at all clear which policy would prevail, particularly since Emkay's application did not snub the River Plan entirely. It proposed to maintain part of the existing gravel pond and to enhance it for recreational fishing and for fish and wildlife habitat.[45] The company also hired an aquatic biologist to consult with it on the details of the mitigation work, which required a 1:1 replacement ratio for any displaced class A lands.[46] Still, despite the proposal clearly contradicting the River Plan's intent, various city staff and decision makers looked for ways to justify recommending approval of Emkay's plan. To gain approval from FEMA for altering the floodway boundary, the developer sent data from its own "computer analysis [which more] accurately represent[ed] the present river flow." To justify the proposed parking lots on class A lands in the floodplain, the applicant suggested that they could claim the 680 parking spots as "supporting" recreation—an allowed use in the floodway. To accomplish this, the applicant zeroed in on various uses allowed along the Greenbelt in the floodway, among which were hunting and fishing. Parking, they argued, was needed to provide access to such activities, and the mall parking could be designated as such.[47] Although Boise's planning staff was persuaded by this argument, members of the Boise Parks and Recreation Commission disagreed, noting that 680 spaces was excessive and could not be justified. Eventually, staff recommended not a denial of the application, but a requirement that the applicant move the parking or ask the city to amend the Metro Plan to permit it.[48]

In early 1986 the city held hearings on both applications, hearing from citizens concerned about river preservation and others who wanted the jobs and economic development the proposal promised. At a Planning and Zoning Commission hearing in January, the applicant's representative pointed to the River Plan's discrepancies when it came to lands like the subject parcel: "The problems we are having with the existing River Plan is there is no implementing ordinance in place to deal with how to mitigate some of the problems. Under the River Plan provisions, gravel extraction ponds are classified 'B,' industrial areas are classified 'C.' Under either of those classifications we wouldn't be here taking [*sic*] about this from a parking issue. [But this parcel] is improperly classified. It is included in Class 'A' [only] because of the floodway definition that occurs on the site."[49]

In the applicant's mind, two problems existed: first, it was hard to conceive of the land as high quality (class A) because its industrial history had rendered the parcel unsightly. Second, the applicant could point to the River Plan's discrepancies and suggest that the city's intent to establish a regional shopping center (which might be allowed on class B or C lands) should trump the conflicting Boise River Plan. Citizens, on the other hand, argued that "encroachment of the Boise River floodway for surface parking is not acceptable to us. . . . The Boise River Plan has not yet celebrated its one-year anniversary as an adopted portion of the comprehensive plan. We believe your support of its preservation of class A habitat, including that found in the Boise River floodway, will provide assurance to our neighbors and alert developers to the strong community support for the Boise River Plan."[50] How the city decided this application was expected to set a precedent for how it would protect the river into the future.

Despite the plan's lack of clarity, the city ultimately approved both the rezone and a conceptual approval for the mall and found that "the requirements of the River Plan may be met on this site."[51] The staff report for the mall concluded that "the proposed pond and pond area enhancement meets the intent of the River Plan regarding 1 to 1 restoration," and planning commissioner questions during the hearing made it clear that the applicant had persuaded the decision makers. No one was able to envision this old industrial land as a "high quality habitat area"; it was designated class A "merely" by virtue of being in the floodway.[52] Once approved, however, the applicant was required to begin shopping center construction within two years, after which time the zone would revert to its original classification.

The record is silent as to why the developer did not immediately begin construction, but that decision killed the project. Another mall—farther from the downtown core—was approved and built in the interim, and by 1989, when the applicant revived the proposal, the city had taken action to clarify important River Plan uncertainties since the 1985 effort. First, federal policies regarding wetlands had changed, making it more difficult to simply fill the gravel pits and pave over them.[53] Second, city engineers had gained a better understanding regarding construction in the floodway and the negative implication for river flows and, therefore, for velocity and erosion. Third, and probably most significant, the city had become stricter in its adherence to the River Plan's policies, as citizen support for resource protection increased. So, after all the controversy, the mall was never built.

The city's clumsy maneuvers with the mall proposal illustrated just how difficult it was to balance economic development, river recreation, and resource conservation. Enticing Boise's population to the river for community building (Greenbelt) and economic development (mall) was an innovative and unique way to deal with the larger trends of deindustrialization and the environmental movement; but achieving those goals while simultaneously protecting the resource through the River Plan was a challenge. The expanding Greenbelt stood as an open invitation, luring humans to the water's edge in pursuit of recreation. Such success stoked human demand for excessive access to the river in the form of whitewater parks and man-made waves, what you'll see today if you stand on the site of the proposed mall. How would a mall have fit in with that plan, and how would a mall affect the ecology of the system? The battle demonstrated the complexities of urban river management, and it was only by the luck of time passing that the proposal died and the parcel was available to become part of a chain of parks thirty years later.

In an event of historical irony, the cleanup, preservation, and reimagining of the Boise River grew out of Boise's unique urban renewal story and the strange coincidence of the city's belated adoption of formal city planning. Boise's own implementation of urban renewal in the early 1970s tore down much of the small downtown's fabric, as it did in cities across the country. The difference in Boise, however, was that rather than implementing urban renewal within the framework of an old city plan and an entrenched urban identity, its leaders were instead working to implement a new city plan, one that placed natural amenities—including the river—at the center of its new identity. Federal adoption of clean water and other environmental laws in the early 1970s nudged the city along in the implementation of its vision. As a result, Boise led a group of small cities in the West in recognizing that the relatively clean river that emerged from enforcement of these laws provided an opportunity to market a new identity for the city.

These two events—urban renewal and the adoption of comprehensive urban planning—converged with Boise's transition from a resource-extractive economy to service-industry future. The city's evolving urban identity was a boon to Boise's new type of worker, who arrived in the city pregnant with ambition and ideas and seeking a particular quality of life, one where recreation and outdoor accessibility were at the center. But Boise's success did not necessarily mean success for the river itself, which became trimmed at the edges, intensively used by humans, and exploited in new and unexpected ways. Of all the cities and towns that line the banks of the Boise River, Boise is renowned for making construction along the river the hardest, thanks to the River Plan and the mall's failure.[54] But true restoration of the river remains in its future. Balancing its wildly successful new identity with the best interest of the resource will be Boise's twenty-first-century challenge.

NOTES

1. "Boise's National Recognition," Build Idaho, accessed August 31, 2021, https://www.buildidaho.com/blog/boise-national-recognition-from-september-2007/.
2. "The 10 Best Cities in the U.S. to Move to Right Now," Curbed, February 10, 2020, https://archive.curbed.com/2020/2/11/21126341/best-cities-to-live-united-states.

3. Census data from 2020 showed Idaho as the second-fastest growing state in the country. Idaho residents from other parts of the state accounted for the largest percentage of in-migration to the Boise Metropolitan Service Area, followed by California, Washington, and Oregon. The COVID-19 pandemic accelerated these trends; California sent 27.5 percent more of its people to Boise in 2020 than in 2019. "COVID-19 Impact on Resident Migration Patterns," CBRE, accessed August 31, 2021, https://www.cbre.com/insights/reports/covid-19-impact-on-migration-patterns; Rachel Spacek, "Idaho 2nd Fastest-Growing State in Latest Census," *Idaho Press*, April 26, 2021.
4. Boise was relatively unconstrained by the "path dependence" described by Scott Page in 2006, a phenomenon that places history at the crux of later outcomes. In the case of legacy cities in the Rust Belt, leaders' decisions over time created institutions and cultures that complemented the economic base upon which the cities depended and without which the cities crumbled. Consider a city like Gary, Indiana, where the steel industry was the economic driver, and the city grew and developed around it. Boise was freer to pursue a new identity away from previously adhered to ideas and cultural systems, since no similar institutional entrenchments existed, and no comprehensive plan had articulated an urban vision before 1964. Page, "Path Dependence," 88, 113.
5. For an excellent review of this period of Boise River development, see Stacy, *When the River Rises*.
6. Considered one of the nation's most polluted rivers during the twentieth century, Ohio's Cuyahoga River infamously caught fire repeatedly. The 1969 fire helped spur passage of the 1972 Clean Water Act.
7. Gottlieb, *Reinventing Los Angeles*, 135–72; Colocousis, "It Was Tourism Repellent," 756–76; Stradling and Stradling, "Perceptions of the Burning River," 515–35.
8. Jenny Price, "Thirteen Ways of Seeing Nature in LA," *Believer Magazine*, April 1, 2006.
9. Stevens, "Feminizing the Urban West," 175–232.
10. Boise Planning and Development Committee, "Development Guide for the Central District of Boise, Idaho," 2, 19.
11. Urban Land Institute, *Boise, Idaho, Capital City*, 14.
12. There is a large body of literature on the history of urban renewal and, in particular, on the effects of deindustrialization on cities. However, the vast majority of the scholarship focuses on Rust Belt cities. Robert Self's essay on Oakland, California, comes closest to providing a framework that is useful for telling Boise's story. In his essay, he seeks "to understand deindustrialization as a product of local planning, civic boosterism, and politics as much as of national macroeconomic processes and broad patterns of capital mobility" ("California's Industrial Garden," 159–80). See Mallach, "Transformation Is Messy Work," 168–88; Tumber, "Fields, Factories,

and Workshops," 224–41; Mallach, "Parallel Histories, Diverging Trajectories," 123–45; Neuman, *Remaking the Rust Belt*; Self, "California's Industrial Garden," 159–80.

13. Village of Euclid, Ohio, et al. v. Ambler Realty Company, 272 U.S. 365 1926; Boise Planning and Development Committee, "Development Guide for the Central District of Boise, Idaho," 22.
14. Idaho Department of Parks, "Boise River Greenbelt Acquisition," sections 3, 11.
15. Idaho Department of Parks, "Boise River Greenbelt Acquisition," sections 3, 11–12.
16. "Greenbelt Needs Board," *Idaho Statesman*, August 20, 1969, 4.
17. Idaho Department of Parks, "Boise River Greenbelt Acquisition," sections 3, 14.
18. Rome, *The Bulldozer in the Countryside*. In his landmark book, Rome describes the impacts of eastern suburban sprawl on the natural environment, arguing that the horrors of experiencing soap emerging from a water tap helped to galvanize the environmental movement. While Boise's growth victimized some of the same natural spaces as those described by Rome—including hillsides and wetlands—Boise's experience was less linear, with leaders allowing such development while simultaneously working to preserve its river. Economic growth depended on the cultivation of this new urban character.
19. "Candidates for Mayor, Council Offer Statements," *Idaho Statesman*, October 26, 1969.
20. "Candidates for Mayor, Council Offer Statements"; "Boise Mayor Retains Lead in Early Vote," *Idaho Statesman*, November 5, 1969.
21. Hayley Harding, "From 'A Great Dream' to a Boise Landmark: The Greenbelt Turns 50," *Idaho Statesman*, September 9, 2019.
22. Ray Smelek, "Part 7: Building a Printer Business," HP Memory Project, accessed June 22, 2019, https://www.hpmemoryproject.org/timeline/ray_smelek/making_my_own_luck_00.htm.
23. Smelek, "Part 7: Building a Printer Business."
24. David Staats, "How an Executive's Hair Dryer Saved Memory Chips . . . Tales of Micron's 40 Years," *Idaho Statesman*, October 18, 2018.
25. L. J. Davis, "Tearing Down Boise," *Harper's Magazine*, November 1974, 32–45.
26. Davis, "Tearing Down Boise," 40.
27. U.S. Department of Commerce, "1980 Census of Population"; U.S. Department of Commerce, "1990 Census of Population and Housing—Idaho"; U.S. Department of Commerce, "2000 Census of Population and Housing—Idaho"; U.S. Department of Commerce, "2010 Census of Population and Housing—Idaho." Some of this was due to annexation of new lands into the city, but the upward trend remains true.
28. U.S. Department of Commerce, "2000 Census of Population and Housing—Idaho"; "2020 Population and Housing State Data," U.S. Census Bureau, August 12, 2021,

https://www.census.gov/library/visualizations/interactive/2020-population-and-housing-state-data.html.

29. Harding, "From 'A Great Dream.'"
30. "1087 W River Street, Parcel R2887290350," Ada County Assessor Land Records/GIS, accessed August 31, 2021, https://adacountyassessor.org/propsys/getCommercialCharacteristics.do?yearParcel=2021r2887290350; "1101 W River Street, Boise, Parcel R2887290275," Ada County Assessor Land Records/GIS, accessed August 31, 2021, https://adacountyassessor.org/propsys/getCommercialCharacteristics.do?yearParcel=2021r2887290275.
31. "500 E. Baybrook Ct., Boise, Parcel R6907050010," Ada County Assessor Land Records/GIS, accessed August 31, 2021, https://adacountyassessor.org/propsys/getCommercialCharacteristics.do?yearParcel=2021r6907050010; Timothy Noah, "The River that Runs through Boise Runs Clear Once Again," *Wall Street Journal*, April 22, 1994.
32. For more on this trend, see Hyra, "The Back-to-the-City Movement"; Glaeser and Shapiro, "Urban Growth in the 1990s"; Kasarda, Appold, and Sweeny, "Central-City and Suburban Migration Patterns"; Laska and Spain, *Back to the City*; Sturtevant and Jung, "Are We Moving Back to the City?"
33. Anderson and Leal, *Enviro-Capitalists*, 12.
34. Stevens, "Feminizing the Urban West," 175–232; see Rome, *The Bulldozer in the Countryside*.
35. Anderson and Leal, *Enviro-Capitalists*, 14.
36. "Chapter 11–16: Boise River System Ordinance," Boise Municipal Code, accessed November 17, 2023, https://codelibrary.amlegal.com/codes/boiseid/latest/boise_id/0-0-0-31831.
37. "Chapter 11–16: Boise River System Ordinance."
38. "Chapter 11–16: Boise River System Ordinance."
39. Boise has taken to naming its parks after women, considering the parks the "ribbon of jewels" throughout the city. "Ribbon of Jewels," City of Boise, accessed August 31, 2021, https://www.cityofboise.org/departments/parks-and-recreation/ribbon-of-jewels/.
40. "Douglas W. McKibben, Area Manager, to Ms. Susan Stacy, Boise Community Planning & Development Dept.," October 8, 1985, CUP; "CUP-153-85/Emkay Development Co., Inc.," 7, CUP.
41. Boise City Planning Department, "Rezoning—Staff Report, File Number: RZ-14-85," 16, RZ.
42. A floodway is the land adjacent to a waterway that is reserved for the discharge of floodwaters. A floodplain, which includes the floodway within it, is more expansive and includes additional lands on the fringe of the floodway.

43. Boise City Planning Department, "Rezoning—Staff Report, File Number: RZ-14-85," 15, RZ.
44. "Township No. 3 North, Range No. 2 East, Boise Meridian."
45. Boise City Planning Department, "Rezoning—Staff Report, File Number: RZ-14-85," 6, RZ.
46. "F. C. Kelley, PhD, Consulting Aquatic Biologist, to R. C. La Fevre, Principal Engineer, M-K Engineers, Inc.," December 3, 1985, RZ.
47. "Douglas W. McKibben, Area Manager, to Ms. Susan Stacy, Director of Planning," January 6, 1985, CUP.
48. "Carla L. Olson, Boise City Community Planning and Development, to Mr. Douglas McKibben, Emkay Development Co., Inc.," January 22, 1986, CUP.
49. "CU-153-85/Emkay Development Co., Inc.," 9, CUP.
50. "Mary Jensen, Vice President of East End Neighborhood Association, to Commissioners," January 28, 1986, CUP.
51. "Carla L. Olson, Boise City Community Planning and Development, to Mr. Doug McKibben, Emkay Development Co., Inc.," January 1, 1986, CUP; "Carla L. Olson, Boise City Community Planning and Development, to Mr. Doug McKibben, Emkay Development Co., Inc.," February 5, 1986, RZ.
52. "CU-153-85/Emkay Development Co., Inc.," 8, CUP.
53. "Carla Martin, Planning Department, to Mayor and Council," February 7, 1990, CUP.
54. Margaret Carmel, "The Boise River: Nature, Development, and Water Quality Shape Its Future," BoiseDev, March 1, 2021, https://boisedev.com/2021/12/22/icymi-2021-the-boise-river-nature-development-and-water-quality-shape-its-future/.

BIBLIOGRAPHY

Archives/Manuscript Collections

Boise Planning and Development Committee. Boise State University Special Collections and Archives, Boise ID.

Capital City Development Corporation. Papers. Idaho State Record Center, Boise ID.

CUP. Conditional Use Permit Application File CU-153-85. Emkey Development Corporation, Inc., Boise City Clerk, Boise City Hall, Boise ID.

Idaho Department of Parks. "Boise River Greenbelt Acquisition," 1967, MSS 006, box 157, folder 19—Sherman Perry. Boise State University Special Collections and Archives, Boise ID.

Perry, Sherman. Papers. MSS 006. Boise State University Special Collections and Archives, Boise ID.

RZ. ReZone Application File RZ-14-85. Boise City Clerk, Boise City Hall, Boise ID.

Published Sources

Anderson, Terry L., and Donald R. Leal. *Enviro-Capitalists: Doing Good While Doing Well.* Plymouth UK: Rowman & Littlefield Publishers, 1997.

Colocousis, Chris R. "'It Was Tourism Repellent, That's What We Were Spraying': Natural Amenities, Environmental Stigma, and Redevelopment in a Postindustrial Mill Town." *Sociological Forum* 27, no. 3 (September 2012): 756–76.

Glaeser, E. L., and J. M. Shapiro. "Urban Growth in the 1990s: Is City Living Back?" *Journal of Regional Science* 43, no. 1 (2003): 139–65.

Gottlieb, Robert. *Reinventing Los Angeles: Nature and Community in a Global City.* Cambridge: Massachusetts Institute of Technology, 2007.

Hyra, Derek. "The Back-to-the-City Movement: Neighbourhood Redevelopment and Processes of Political and Cultural Displacement." *Urban Studies* 52, no. 10 (August 2015): 1753–73.

Kasarda, J. D., S. J. Appold, and S. H. Sweeny. "Central-City and Suburban Migration Patterns: Is a Turnaround on the Horizon?" *Housing Policy Debate* 8, no. 2 (1997): 307–58.

Laska, S. B., and D. Spain, eds. *Back to the City: Issues in Neighborhood Renovation.* New York: Pergamon Press, 1980.

Mallach, Alan. "Parallel Histories, Diverging Trajectories: Resilience in Small Industrial Cities." In *Revitalizing American Cities*, edited by Susan M. Wachter and Kimberly A. Zeuli, 123–45. Philadelphia: University of Pennsylvania Press, 2014.

———. "Transformation Is Messy Work: The Complex Challenge of Spatial Reconfiguration in America's Legacy Cities." In *Revitalizing American Cities*, edited by Susan M. Wachter and Kimberly A. Zeuli, 168–88. Philadelphia: University of Pennsylvania Press, 2014.

Neuman, Tracy. *Remaking the Rust Belt: The Postindustrial Transformation of North America.* Philadelphia: University of Pennsylvania Press, 2016.

Page, Scott E. "Path Dependence." *Quarterly Journal of Political Science* 1 (2006): 87–115.

Rome, Adam. *The Bulldozer in the Countryside: Suburban Sprawl and the Rise of American Environmentalism.* Cambridge UK: Cambridge University Press, 2001.

Self, Robert O. "California's Industrial Garden: Oakland and the East Bay in the Age of Deindustrialization." In *Beyond the Ruins: The Meanings of Deindustrialization*, edited by Jefferson Cowie and Joseph Heathcott, 159–80. Cornell: Cornell University Press, 2003.

Stacy, Susan. *When the River Rises: Flood Control on the Boise River, 1943–1985.* Denver: University of Colorado Press, 1993.

Stevens, Jennifer. "Feminizing the Urban West: Green Cities and Open Space in the Postwar Era, 1950–2000." PhD diss., University of California, Davis, 2008.

Stradling, David, and Richard Stradling. "Perceptions of the Burning River: Deindustrialization and Cleveland's Cuyahoga River." *Environmental History* 13, no. 3 (July 2008): 515–35.

Sturtevant, L. A., and Y. J. Jung. "Are We Moving Back to the City? Examining Residential Mobility in the Washington, DC Metropolitan Area." *Growth and Change* 42, no. 1 (2011): 48–71.

"Township No. 3 North, Range No. 2 East, Boise Meridian." Original Survey. Boise ID: U.S. General Land Office, 1868.

Tumber, Catherine. "Fields, Factories, and Workshops: Green Economic Development on the Smaller-Metro Scale." In *Revitalizing American Cities*, edited by Susan M. Wachter and Kimberly A. Zeuli, 224–41. Philadelphia: University of Pennsylvania Press, 2014.

Urban Land Institute. *Boise, Idaho, Capital City. A Report to the Boise Planning and Development Committee, Inc.* Washington DC: Boise State University Special Collections and Archives, 1966.

U.S. Department of Commerce. "1980 Census of Population." Washington DC: U.S. Census Bureau, 1983.

———. "1990 Census of Population and Housing—Idaho." Washington DC: U.S. Census Bureau, 1990.

———. "2000 Census of Population and Housing—Idaho." Washington DC: U.S. Census Bureau, 2003.

———. "2010 Census of Population and Housing—Idaho." Washington DC: U.S. Census Bureau, 2012.

6 Urban Ecology

Rewilding Rivers and Creating Outdoor Classrooms in Milwaukee

Christian C. Young

Of all the days I have spent along the rivers in the city of Milwaukee, identifying trees and watching birds, one of my most memorable mornings involved carting saplings down to a floodplain overgrown with invasive grass. With a group of volunteers that included my immediate family, we dug holes, put screening around the small trees to protect them from nibbling deer or rabbits, and gave their newly transplanted roots a fresh drink of water and thick coating of mulch. The screened saplings looked ready for a season of growth. I imagined coming back in future years to see them mature into a new riverside forest.

The work of that day provided its own reward, learning a few new requirements for the trees, spending time with my family and members of our larger community, and enjoying the sunshine in a quiet spot by the river. A few months after we planted those saplings, floodwater washed over them. Runoff from a week of torrential rains pulled away the screens, bent the stakes, and buried many of the plantings in silt. Return visits have revealed not a single survivor from our efforts, but in subsequent years, other volunteers have planted again, and saplings are taking hold.

Creating the conditions for this sort of community effort requires a long view of rewilding. At the core of that vision is a belief that people in cities, like anywhere else, need to maintain a connection to nature. Having places nearby where neighbors can gather, children can explore, and wild things can thrive makes it possible to connect to nature. A river corridor offers a distinctive opportunity, as people in Milwaukee have recognized. Numerous organizations and government agencies share the credit for successes thus far, yet there is a lot of work left to do.

One group, the Urban Ecology Center (UEC), has explicitly embraced the mission of connecting people in cities to nature and each other. The UEC serves as an environmental education community center to organize individuals, other nonprofits, government-led initiatives, and community conservation efforts around that mission. Through its land stewardship program, the UEC has coordinated tree planting near the rivers along with efforts to establish and activate outdoor classrooms in urban places. The story of rewilding river corridors in Milwaukee can be told through the lens of economic revitalization or environmental remediation, but at its core, the greatest success has been in restoring sections of nature to the community's heart. The practical work of the UEC over the past twenty-five years has served as a means for transforming the city.[1]

As a concept, rewilding reshapes landscapes as wide as the imagination and connects smaller preserves with robust habitat links.[2] Elaborate rewilding schemes require policy change at national scales with global cooperation. Such broad support to protect the environment, however, requires local efforts that foster a sense of stewardship for small places close to home. People in cities too often live with almost no connection to even the small, enduring patches of nature. Children and young people rarely experience days of rambling and exploring that characterized the early days of so many enthusiastic natural scientists and environmental advocates. If rewilding is ever to succeed as a continental priority, we need to see its potential in schoolyards, within vacant lots, and along riverways.[3] Multiple nonprofit organizations can partner with municipal governments to establish outdoor education programs tied to protecting urban green spaces and water resources, first in one neighborhood, then another. This small-scale, local approach can be instrumental in rewilding efforts on key sections of a city's parks and rivers.

The names of the rivers flowing through the heart of Milwaukee do not appear on the lists of great transport routes across the United States. They are not described as scenic in venerable works on the environment in the traditions of Edward Abbey or Wallace Stegner. The rivers also have not famously caught fire during the city's history. Through decades of reshaping and quietly disappearing from view, they were transformed into drainage ditches and stagnant canals connecting sections of the harbor. Recently, however, the rivers have helped to fuel a revitalization in urban life and connections to nature. The Milwaukee, Menomonee, and Kinnickinnic Rivers converge in the city's inner harbor, a global port on Lake Michigan in southeastern Wisconsin.

A Distinctive but Familiar Story of Decline

Milwaukee's thirst for clean water matched and exceeded the seemingly boundless supply beginning in the 1870s. Tanneries and breweries occupied and reshaped the riverbanks in the early days of non-Indigenous settlement. Agricultural and forest products made their way into the Milwaukee Harbor of Lake Michigan on a scale that rivaled Chicago for a time. Milwaukee's growth disrupted the ecosystems of its rivers, their surrounding wetlands, and lakefront. As with many other industrial cities in the region, the city's urban rivers were severely degraded and even considered dead. Long stretches of it were confined to concrete-lined channels, bound by steel piling and stilled behind dams.[4] As in most American cities where industry once boomed, Milwaukee experienced dramatic environmental degradations along its waterways. Population surged through the middle of the nineteenth century as machine shops, leather tanneries, and meatpacking houses, along with the signature breweries, expanded. This growth ushered in the despoliation of the landscape, particularly on the city's harbor and near watersheds across generations of development. The real threat to Milwaukee's waterways began in the 1870s, when almost seventy-five miles of sewer pipe were installed to carry waste away from homes and factories, emptying into the nearest river.[5] Rather than address the pollution at its doorstep, the city built intake lines farther out in the lake to get pure water.

The same impulse to avoid the growing city's pollution problems by extending services farther from its core led city leaders and entrepreneurs to look upstream along the Milwaukee River. The opportunity for recreation and relief from the city's congestion and grit emerged in new locations across multiple generations within a few miles north of downtown. On those sections of river, dams have come and gone, along with flour mills and industrial operations, but also swimming schools, winter skating parks

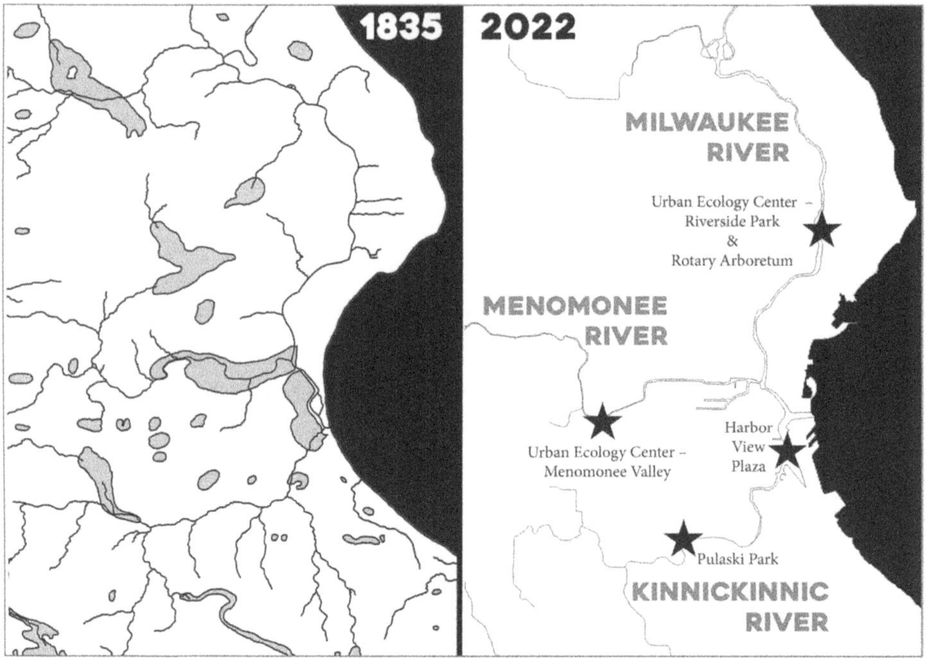

7. Map of the Milwaukee watershed. On the left are historic wetlands and watercourses from public land survey maps from 1835 to 1837 held by Wisconsin Commissioner of Public Lands. Map on the right shows contemporary watercourses with points of interest added by the author. Courtesy of Reflo and Milwaukee Metropolitan Sewerage District (https://refloh2o.com/milwaukee-community-map).

and ski jumps, boathouses and rowing clubs, beer gardens, and amusement parks. Some of these endeavors flourished seasonally for a year or two, and others endured for a decade or more. Memories from the "In-Town Up North" locations have faded. Photographs in historical archives made their way into books about the neighborhoods that attracted families who could afford to escape the increasing congestion and pollution of the city.[6] Fifty years of development along the river are compressed into the blink of an eye, glancing at grainy photos and a few jerky film clips of swimmers, boaters, and skaters. By the 1920s most of the land between the river bluffs had been abandoned. Railroad lines cut off the neighborhoods from the water, and people learned to avoid the increasingly polluted water, especially as more factories went up.

The degradation of Milwaukee's harbor continued upstream along the Kinnickinnic and Menomonee Rivers. The waters of Lake Michigan rise and fall in response to prevailing winds across the lake, which can shift river currents and carry water upstream in irregular patterns across flat water near the harbor. Early efforts to force putrid water more rapidly from the Menomonee out to the lake were abandoned by the 1880s because even the ability to pump as many as seventy million gallons per day out of the river did not provide a solution. Additional spending to expand this approach would be prohibitive. The city tried a different option to solve the sluggish flow, building flushing stations, which pumped fresh water from the lake into upriver sections to accelerate the flow. This was practical in the Milwaukee River, situated for a portion of its course parallel to the lakeshore. Pouring clean water into the river improved the situation. A second station opened in the Kinnickinnic River. These flushing stations remained in operation for decades but ultimately could not keep pace with the growing city and its reliance on dumping waste of all kinds into the waterways. The rivers continued to decline without any meaningful remediation for more than half a century.

Planning and Implementing Watershed Improvements

The Southeastern Wisconsin Regional Planning Commission (SEWRPC) took shape in 1960 in response to decades of uncoordinated development across many jurisdictions. The fact that the commission originated when it did and has persisted provides evidence of forward-looking leadership of the region's municipalities, at least as a response to pollution concerns.[7] SEWRPC soon became recognized by the pronunciation of its acronym, "sewer-pack," and its jurisdiction extended across eight watersheds of varying quality. The channelized urban stretches built for the Kinnickinnic River contrasted with rural wetland habitat that supported abundant plants and wildlife, especially in the Menomonee and Milwaukee River basins. Over time, streams in those areas became polluted by agricultural runoff, adding to the challenge of determining how to prioritize the growing problem. The combined system of storm and wastewater sewers built in the nineteenth century could not consistently handle the volume of the twentieth-century city. Two treatment plants near the lake would reach capacity about fifty times per year on average, resulting in diluted sewage overflows into the river system and lake, decade after decade.[8] The Milwaukee River itself overflowed during twenty-four flood events between 1915 and 1968, causing damage in neighborhoods constructed without deliberate planning prior to SEWRPC. Flooding threats continued even as the commission began to assess the options for better infrastructure. The plans eventually included comprehensive efforts to address flooding and pollution through zoning for land use in flood plains, the development of coordinated park space, and management of upstream lakes. These regional efforts made possible the modest recovery of urban rivers, but SEWRPC offered no plans that could be considered a step toward rewilding.

Rising awareness of the need for regulation and coordinated watershed management reached a turning point across the country as well. The sight and smell of stagnant surface waters became an affront to emerging environmental sensibilities. Earth Day reflected the shift, and its champion was Sen. Gaylord Nelson (WI). The message resonated from ocean beaches and snowcapped mountains to roadside ditches and stinking rivers. In Cleveland the Cuyahoga River famously caught fire in 1969, but oil slicks and other flammable materials had been burning on its surface for a century. Congress passed the Clean Water Act in 1972, formalizing a national, environmental agenda and providing resources for change. People in cities had begun to take new notice of the causes and need for action, and the focus was on human health. If clean water was to be a priority, the focus would be on cleaning up pollution and providing pure drinking water. If anything, interest in restoring wild places dropped even further from view in industrial cities. Milwaukee was no exception, especially after struggling for decades to make progress on its health problems.

The earliest citywide sewerage commission in Milwaukee had been established in 1913 with a narrow focus on human health. Over time, the city and county expanded treatment plants as a way of keeping the waste of a growing metropolis out of Lake Michigan. More generally, however, early efforts did little to protect the environment. The failure to reduce and remediate pollution attracted wider attention. In the 1970s a lawsuit from the state of Illinois took the issue to the U.S. Supreme Court before an agreement reached with the Wisconsin Department of Natural Resources set oversight of Milwaukee's waterways on a new course. The creation of the Milwaukee Metropolitan Sewerage District (MMSD) in 1982 formed the backbone of an infrastructure plan to address storage, treatment, and disposal methods for wastewater.[9]

When construction of an expanded, combined sewer storage system began in 1985, the goal of eliminating overflows into the lake finally seemed achievable. It took until 2010 to complete the system, which consisted of more than twenty-six miles of tunnels hundreds of feet below the surface. The city finally had the capacity to treat all of its domestic, industrial, and stormwater runoff, except in the rare case of extremely heavy rainfall when the system was already nearly full.[10] This progress, though long overdue, opened new space to envision better uses for the city's surface water, which still occasionally conjured images of swimming schools, beer gardens, and amusement parks. An expanded ecological sensibility about wildlife habitat and restoration of plant communities informed that vision. The notion of rewilding rivers in Milwaukee began to weave into the city's ambition for revitalization.

Case Study: Transformations along the Milwaukee River

Industrial cities like Milwaukee declined dramatically due to broader economic shifts, even as environmental norms evolved. In many places, efforts to "clean up the city" offered a promising alignment. A community effort took shape: bring downtown Milwaukee back to life by focusing on its riverfront. Mayor John Norquist led an initiative to establish the Milwaukee RiverWalk when he was elected in 1988. He had seen the San Antonio River Walk during his training days in the Army Reserve, and like city leaders elsewhere, Norquist and others had been inspired by public access and walkways along rivers in London and Vienna. The vision had a pragmatic focus in Milwaukee.[11] As the economy struggled in the preceding decades, the most offensive sources of industrial pollution had shut down. The city's population declined, and jobs along the river corridor disappeared. Mayor Norquist found allies among property owners who saw the larger project as a collective investment opportunity. Together, they believed that municipal investment in stretches of land along the river could drive market demand for waterfront property. That demand, in turn, could be leveraged to pay for public works that would add vitality to the city. Many other cities tried the same general approach. Norquist's vision, over the long haul, has held up in Milwaukee.[12] City hall required that a public walkway along the Milwaukee River be fully accessible and connected, block by block, to the streets and bridges of the city. The same vision of municipal leadership fueled the creation of a greenway project to encompass a stretch of the river still visible in the grainy photographs of "In-Town Up North."

The Milwaukee River Greenway emerged in novel form due to its distinctive topography and history, along with public policy informed by a vision of connecting people to nature in this section of the city.[13] The river corridor contained the ruins of recreation facilities of a bygone era, and it had been largely abandoned for decades. A dam separated this stagnant section of the river from the harbor below. On the bluffs above, a series of parks served their respective neighborhoods in relative isolation. Few people ventured down toward the water, where tangled paths overgrown with invasive species slowed progress. Arriving at the slow-moving river presented an unwelcome reward—a polluted slough rather than a sparkling lake.

Riverside Park had been a gem in the Milwaukee County Parks system of green space, originally designed by Frederick Law Olmstead. As in other cities, Olmstead planned an "emerald necklace" of public parks. Connected to another park on the shore of Lake Michigan by a well-adorned parkway, Riverside had attracted neighbors to a vibrant section of river for decades. When the water conditions declined, so did the park. Residents of the affluent East Side neighborhood around its borders began responding to fears about crime in the park. Distress about criminal activities matched civic dissatisfaction with the untamed and unkempt condition of the wooded area along the river. Police intervention seemed as unlikely to provide a long-term solution as increased park maintenance, given budget constraints on both the city and county. Addressing the ecological damage would be an even lower priority. Starting with environmental education in such a place was not an obvious next step.

A visionary scientist and educator, Else Ankel, evaluated the situation from the perspective of a neighbor who wanted to transform the environment at every level. She recognized the potential of the woods, open space, and trails down to the river as an outdoor classroom in need of care. Ankel had retired and brought together community advocates who shared her vision for healing the land. By 1998 the group decided to hire someone to lead what they called the Riverside Park Environmental Center.

8. Hikers on the Milwaukee River trail near the Urban Ecology Center. Courtesy of Urban Ecology Center.

When Ken Leinbach came on board as the director, he brought training as a science teacher and experience running summer camps and outdoor learning programs. He added a broader vision to the locally focused neighborhood group, which began to operate classes and programs out of a trailer. The vision included an ambition to expand their programs. Eventually, he hoped, every child would have contact with nature and a mentor to guide them, every day, across Milwaukee and in other cities, as well. The ambition was rebranded as the Urban Ecology Center. They hired a second teacher and began partnering with nearby schools to provide regular field trips throughout the school year that brought students back to the park again and again.

The UEC became (and remains) an environmental education community organization with a mission "to connect people in cities to nature and each other." Notably, the original ambition to bring children into Riverside Park quickly expanded to include efforts to provide lessons in restoring the wild spaces along the river. Offering students an experience in nature, UEC educators showed them which plants were harmful to the ecosystem and which would provide habitat for native wildlife. Land stewardship efforts to remove invasive species began to attract volunteers in large numbers. A community science program to monitor and study the impact of native planting and habitat restoration could demonstrate the ecological value of the work. Within five years, educational programs with schools were flourishing, volunteers enthusiastically arrived daily, and funders wanted to provide the means of expanding the work through a capital campaign. Along the way, the UEC became a hub for rewilding a key section of the city's river corridor in the ongoing revitalization of the urban scene. The UEC opened a new, state-of-the-art, sustainably built community center in 2004. Leinbach and the growing staff had a location to showcase the transformation that was taking place along the river. The combination of environmental education with land stewardship and community engagement became pillars of this work.[14]

City leaders, who continued to coordinate ongoing efforts to improve conditions closer to downtown, took note of what was happening around Riverside Park. At the same time, the UEC's work seemed to fulfill larger hopes envisioned by Milwaukee's Kiwanis and Rotary clubs. Those clubs had joined forces to establish the River Revitalization Foundation and commissioned a survey for the Wisconsin Department of Natural Resources. That survey had resulted in a 1991 recovery plan and eventually established an "urban land trust" for the corridor. For more than a decade, the recovery plan fluctuated between a grand vision of clean water with restored forests, and a reactive effort to hold off the development of high-rise apartments over the river. Working more directly toward the grand vision of a protected wild space along the river, a group that called themselves the "River Rats" convened. They included Leinbach from the UEC, a historic preservationist who owned property between the UEC and the river, a philanthropist, a staffer from Milwaukee County Parks, a National Park Service employee, and a member of the River Revitalization Foundation.[15]

Without a clear public mandate for restoration of the river, however, the threat of development and encroachment into the riverway continued. In 2005 the land along the river was still mostly neglected—Riverside Park was an exception, and much work remained there, as well. A property owner clearing trees on a section of land sparked action by the River Rats, who helped raise the alarm to the wider community. Within a matter of months, they formed the Milwaukee River Workgroup and published a plan for a central park along the river that would be larger than Olmsted's New York creation. The workgroup kept up pressure to establish a zoning restriction to limit any new construction that would be visible from the river. This "viewshed" became the basis for an overlay district, adopted by the city of Shorewood in 2009. Shorewood had jurisdiction over the northern section of the district. The city of Milwaukee gave approval the next year, and the Milwaukee River Greenway Coalition was formed to carry out additional planning.

Throughout this period of slow but steady progress, the UEC used environmental education as a rationale for land stewardship in the city and as a means of engaging the community. The organization brought more partners and funders to the table.[16] They continued to expand the outdoor classroom of Riverside Park. The Rotary Club of Milwaukee added the forty-acre Centennial Arboretum to the Greenway in 2013. The Milwaukee County Parks system improved access to the river on its property on both sides of the river. The Greenway expanded to protect more than eight hundred acres and provide five miles of connecting trails. Upriver from Estabrook Park in the Greenway, removal of a dam that had fallen into disrepair has helped restore additional river features and reduce maintenance costs along the corridor.[17] The ongoing rewilding of the Milwaukee River resulted from a ripple effect of establishing an outdoor classroom in alignment with city revitalization efforts. These projects have created new wild areas and facilitated the building of trails down to and along the river. The paths connect parks on both sides of the river.[18] As it turns out, this rewilding in Milwaukee was not limited to one river, due in large part to the replication of key collaborations.

Case Study: Transformations in the Menomonee Valley

Urban revitalization and economic development also played a major role in early efforts to bring nature back along the banks of the Menomonee River. And once again, the ambition to create an outdoor classroom provided an additional spark and sustained fuel for wider community engagement in the Menomonee Valley. The valley includes the portion of the river that had once been a vast wetland, confined by low bluffs to the north and south until it flowed into Lake Michigan. City leaders and community advocates could point to the economic potential of improved river conditions throughout the watershed, along with opportunities to improve public health outcomes and neighborhood ties within the city. Residential areas bordering the valley became disconnected from the industrialized valley during the course of the twentieth century, aside from the daily passage of workers from their homes to the manufacturing plants and railyards. Those links dissolved, too, when industry and transportation in the valley declined as the economy shrunk in the 1970s. Land restoration and environmental education possibilities in the valley matched the accomplishments of efforts north of the city along the Milwaukee River.

Upstream the state and MMSD—the local wastewater authority—led efforts to address long-term concerns about reducing the flow of pollution through the region and into Lake Michigan.[19] In the valley, a plan that included a market, engineering, and land-use study was published as part of citywide work in 1998.[20] The Major League Baseball stadium, with parking lots that straddle the river, was rebuilt in 2001. Additional urban planning and revitalization attracted new industry to the vacant land and abandoned factories. Plans to control stormwater included specially planned wetlands and spillover basins. There was little room to reshape the main channel of the river, however. It was bounded by steel piling. Above its banks, parking lots and new construction for companies recently attracted to the valley might have threatened to further imperil the river. Instead, those companies joined a consortium of community organizations, developers, land planners, and government agencies to create a vision that would return both economic and ecological wealth to the area.

Together, these groups formed the Menomonee Valley Partners in 1998. Within a few years, plans took shape to remediate additional sections of land in the valley. North of the river, construction of new light industrial buildings created opportunities to situate small wetland areas between parking lots for improved stormwater capacity. Planting flood-tolerant species native to the area further enhanced the viability of plans to include mixed development, land conservation, and watershed management. A large section of vacant brownfield south of the river—once the site of a manufacturing operation that provided railroad car parts for the nation—remained covered in gravel and enclosed by fencing. The plan for a particular twenty-four-acre site included managing construction fill from a freeway project downtown.[21] Menomonee Valley Partners championed this effort to consolidate and confine waste, to reduce the cost of transporting it farther from the city, and to reinvest in the vision for an urban green space.

As part of this brownfield remediation, the UEC joined the planning process with the vision of building another outdoor classroom. The planned park would feature a landscape that simulated glacial landforms. Using funds that would otherwise be spent hauling waste material, the plan called for establishing a layer of clean soil over the fill, then covering a rewilded space with native prairie plants and strategically placed trees to create areas of oak savannah and riverbank woodland habitat. The plan, an ambitious example, won a national award in 2011 from the American Society of Landscape Architects.[22]

An important component of this plan, the outdoor classroom, resonated with neighbors and collaborators of the Menomonee Valley Partners, since it brought the promise of programming from the UEC.[23] In neighborhoods south of the valley, existing parks were embedded among residential streets. Small blocks of green space offered few amenities there. The possibility of creating a new outdoor classroom in that section of the city evoked a special urgency. The success of the UEC model hinged on finding the potential of degraded land on the edges of neighborhoods, with access that could include aquatic habitat as part of the ecological restoration. The brownfield area in the Menomonee Valley provided opportune space to test the model further. UEC supporters identified a vacant building on the street nearest to a pedestrian access point into the valley and purchased it as the home of a new branch location, the third in the city. With the prospect for ongoing environmental education programming along with land stewardship and community engagement, the potential for this rewilding project seemed assured. The organization played a significant role in the transformation, widening awareness of what could be accomplished and playing its familiar role as a convener of other groups, including MMSD, city and county officials, business leaders, and organizations that now included the Sixteenth Street Community Health Center.[24]

Large sections of the construction site, complete with glacial landforms, were seeded with prairie vegetation. Planting trees on sections of emerging parkland began as soon as earth movers completed the work of reshaping the drainage basins and future wetlands. The design for the main restoration project included an east-west strip of land south of the river and bordered by the remaining rail lines beyond. It extended from the Menomonee Valley branch of the UEC at Thirty-Seventh Street about ten blocks east toward

Lake Michigan. A bridge across the river for bikes and pedestrians was built on the site of an older bridge that had provided access for workers to the valley's factories. Another bridge would connect additional trails to a cul-de-sac used mainly by container trucks that carried away products from the remaining and revitalized industry in the center of the valley. And a high bridge over the railroad tracks linked the restored land for the first time to the county's "domes," the Mitchell Park Conservatory. When a naming contest was launched to dub the resulting park, Three Bridges Park emerged as the obvious choice.

The landscape design brings together the natural, cultural, and historic elements of the space. It includes access to the river on the north side, at the west end of the park, and on the south bank near the middle bridge. A main trail, wide and paved for all users but especially attractive to cyclists, connects to the wider network known as the Oak Leaf Trail.[25] A variety of loops and spurs take hikers and wanderers closer to the river, through remnant forests, and over the mounds of rubble that have been shaped to recall glacial features, like scale-model eskers and moraines. All of this land shaping was completed by the fall of 2012. When the park opened the next year, most of the soil had been planted with rye and other fast-growing native plants that would serve as pioneer species. Additional seed for a wider variety of prairie vegetation included in the mix had time to germinate on schedule and emerge almost unnoticed during the first few growing seasons.

Use of the land as an outdoor classroom began before the planting had even been completed. School children and community members were introduced to the place by environmental educators and land stewards at the UEC. Staff from the research and community science team began leading bird walks as well as documenting species and monitoring for reptiles and amphibians, an ever-increasing variety of insects, and mammals in expanding numbers. From one year to the next, the sections of prairie revealed familiar patterns of diverse and patchy species adapted to the varied soil and moisture conditions resulting from contours of the land. With the design of the glacial forms adding to the diversity of the plant community, areas of more and less direct sunlight revealed greater resilience through the changing seasons. Along the river, some trees in healthy condition remained even after decades of neglect; the design allowed for them to continue to grow, casting long shadows over the water even in the

9. Data collection at the canoe launch in the outdoor classroom along the Menomonee River in Three Bridges Park. Courtesy of Urban Ecology Center.

height of summer. In places where there was little remaining vegetation, or it consisted primarily of invasive species, land stewards cleared the river banks, and native, fast-growing willows, dogwoods, and flood-resistant shrubs were planted using a "livestaking" technique, inserting a fresh-cut branch into the wet soil near the water in the spring. Those branches developed roots into the soil and began to sprout leaves and branches within a few weeks. A decade later, the river was lined with dense, native trees and shrubs that were planted in a matter of days by volunteers.

Restoration efforts along the river and across the prairie did not end with the initial plantings. The UEC used its education programming with school children to promote ongoing work. Volunteers were drawn to the valley in all seasons to monitor bird life, remove invasive species, collect seeds, add new plantings, and maintain the trails. The lessons of this outdoor classroom became increasingly rich. Students were welcomed to the park by UEC educators, typically walking under the railway through a tunnel that has a mural painted to denote the "valley passage." Access to the trails and bridges felt like entering another world. Bringing educational programming to the park and river every day, in every season, has helped students to see the rhythms of nature and recognize the potential for multiple connections with it. They have seen changes, small and large. The prairie has burned in areas, sometimes through planned and carefully controlled land stewardship procedures, and other times as a result of vandalism and arson. Although the consequences of fire have risked damage to the surrounding neighborhood, the prairie thrived after each burn. At all times, teams of city firefighters responded quickly to ensure that the flames were confined between the river and the buffers created by trails, rails, and remaining vacant lots.

The legacy and ongoing potential of the Menomonee Valley have served as an example of what an urban river can become. It was rewilded for the intrinsic value of providing a glimpse of nature in something like its condition prior to the settlement of the city. Perhaps more importantly, it became an outdoor classroom. It emerged as an illustration of how land stewardship can be the shared work of a community. The valley came to offer a refuge for individuals who want to hike or paddle or bike along a river that has provided a home for birdlife and an increasingly robust ecosystem of aquatic and savannah species in close proximity. It became a collaborative product of science and community engagement on many levels.

Toward Rewilding on the Kinnickinnic

Milwaukee's Kinnickinnic River on the south side of the city drains the most densely populated watershed in the state. The neighborhood is highly residential with houses packed onto tiny lots on a grid of streets and alleys laid out in the boom of early twentieth-century urban growth. City planners left very little greenspace in these neighborhoods. Over time, large areas of impervious surface filled in around industrial sites and shopping centers. By the 1960s the need to accelerate stormwater drainage from the congested cityscape led engineers to design a system of concrete-lined channels that transformed almost the entire river and its tributaries into outsized gutters. The river was dropped farther below the street grade and essentially disappeared from view. In the winter, small pools of water would freeze solid. For much of the summer, long stretches of concrete would dry out completely.

Between rains, the sloping concrete served as a broad sidewalk, a shortcut across the river, a place to drop out of view under bridges.

The Kinnickinnic was not entirely forgotten, however. Spring snowmelt and summer thunderstorms converted the uneven trickle at the bottom of those channels into a torrent. Too often, an unsuspecting pet was swept away. The temptation of the raging water attracted children, both curious and adventurous, sometimes with tragic results. The engineers admitted the error of their assumption that the concrete would solve the problem of drainage in tight spaces. In fact, deferred maintenance on the channel compounded over decades, and the cost of repair created a different calculus for how the problem might be addressed by a new generation of engineers and city planners.

Decades after the concrete channel was poured from Forty-Third Street to Sixth Street, converting the Kinnickinnic River back to a more natural flow became an ecological possibility, a social responsibility, and an economic necessity. The river had disappeared, replaced by what amounted to open sewers. The water in the channel stagnated during dry periods and became devoid of all life, other than confined organic matter that stunk and festered. Section by section, amid residential neighborhoods crowded with modest homes, MMSD engineers created a plan to remove the concrete channels. On an impressive scale, the river now bends through an urban park amid wildflowers and a naturalized streambed.[26]

The Medical College of Wisconsin and the nearby Sixteenth Street Community Health Center study the generally recognized benefits to human health of time spent in nature, such as in the Menomonee Valley and along the Kinnickinnic. New evidence continues to serve as a counterpoint to assumptions about how the urban rivers cause infectious disease. The river's new role, in rewilded fashion, improves human health outcomes associated with connecting to nature. The Sixteenth Street Community Health Center pairs environmental health with neighborhood revitalization, stewardship, and education. Programming focuses on children and families to serve the community broadly.[27] On the Kinnickinnic, community programming that supports rewilding more broadly has continued to build. Given the economic resources available to remediate destructive flooding from stormwater and the impact on neighbors of reconstructing sections of the river, funding has flowed into efforts to engage residents who have experienced the worst of the river's past and look forward to a rewilded future.

Since 2015 the effort to rewild the river has attracted attention from a citywide public art program, led by New York-based artist Mary Miss and the City as Living Laboratory project. Miss guided the installation of public art along the Milwaukee RiverWalk downtown in the 1990s.[28] She returned to the city to initiate "Watermarks" in 2017.[29] After riding a borrowed bicycle along the quiet streets near the river between Sixth and Sixteenth Streets, and seeing improvements made by MMSD downstream from that section, she was inspired to include the Kinnickinnic as a significant focus of the citywide watershed project. That project has built on aesthetic awareness of the river's place in the community and a wider call for recognition of how watersheds continue to change in a period of dramatic climate change. Numerous examples of local art were installed along the river in recent years. The first installation in the citywide project, at Acosta School, reflects the engagement of the health clinic, local educators, and the arts community in the neighborhood around the Kinnickinnic River.[30]

Moving Downstream

Individual acts of physical engagement sustain a community's interest and support a relationship with nature. The time and effort put into land stewardship activities—designing, planning, clearing, planting, watering, and pruning, as well as collecting seeds for planting again—are distinctive components of rewilding on the local scale in urban areas.

Recently, beaver activity appeared along the stretch of the Milwaukee River where I worked with other volunteers to plant trees on that sunny morning. The animals find abundant new growth of woody vegetation, including newly planted saplings. They chew the bark off large trees, topple maturing cottonwoods, and strip branches from willows. The trimmings around the ragged stumps provide unmistakable evidence that beavers live nearby, probably in a lodge under an overhanging bank. They will not build a dam across this wide stretch of water, but they have plenty of food and no predators. The soil on this section of the river had waited underwater for many decades and needed remediation. After a few decades of cleanup and restoration to serve as an outdoor classroom and demonstration of nature in the city, the job is not yet done. The trees there seem like a hopeful start.

Now, the beavers have made short work of a decade's hope. They continue to sheer off more of these trees, prompting philosophical questions from land stewards, environmental educators, and others who envision a rewilded river. Are the beavers a sign of success, or the latest confounding factor in establishing a sustainable habitat for a wider variety of plants and animals?

For many people who have worked to establish native vegetation along the river, the flourishing of that vegetation has served to mark the Greenway's success. In the context of the work on Milwaukee's rivers, a larger lesson emerges in these outdoor classrooms. In planting trees on the floodplain, a community learns the patience it takes to expand a forest. The effort moves forward before suffering an unexpected setback. The failures might yield new successes. Thinking of a river in an urban area through this lens can serve to provide deeper insights into the process, expanding the potential for understanding what it means to rewild these places. Ecological learning gained here is more than knowledge of how to eliminate pollution or figuring out what will grow. It is also seeing the system function despite many missing pieces and broken connections, recognizing where more repairs can be made. Deciding what to do with beavers, or whether to do anything at all, deepens our engagement. Over time, we can learn more from community participation in these decisions and our collective response to new challenges. This is a different opportunity than what we find in rural areas with more connections intact, or even wilderness preserves with significant remnant biodiversity. Ultimately, learning in an urban area also has the potential for even greater influence on wider appreciation of nature.

Access to remote natural areas is particularly challenging for members of lower-income communities, especially people of color. The rarity of visits may be all the more inspiring, but the contrast between the beauty of wilderness far from home and the blight all around one's own neighborhood becomes unbearable when combined with other disadvantages faced by those who lack access. More people from diverse backgrounds can visit the outdoor classroom in a city more frequently. Community members can participate in land stewardship on a regular basis. School children can visit multiple times throughout the school year. They can visit with their friends after school and invite their families along on the weekend. The experience of being in nature in the city offers a transformative experience. The process of engaging directly in the restoration of vacant lots and brownfields, as occurred along these rivers in Milwaukee, provides another measure of hope for the participants. It also gives them agency. Rather than sitting back to see how a team of experts might restore a succession of ecological communities, these projects use a citizen-science approach to monitoring, where community members of all ages help to design the studies and volunteer to carry out data collection. This approach results in sharing information regularly and directly with neighbors around the park. Staff and volunteers together celebrate their success as a hallmark of this scientific work. The UEC became the core of these educational, stewardship, and engagement programs. Partnership with many other organizations pointed the way to alternative paths to success in rewilding areas along the rivers. The value of these collaborations cannot be overstated, and the process suggests that there are many other possibilities where rewilding serves as a wider goal for cities with degraded river corridors.

NOTES

1. Van Auken, "Toward a Fusion of Two Lines of Thought," 49; Leinbach, *Urban Ecology*, 8.
2. In popularizing and institutionalizing the concept of rewilding, Dave Foreman placed the highest priority on the widest possible implementation of the idea. He established the priority of "megalinkages" to connect areas on a continental scope. The ambition was, and is, a departure from the island-like focus of nature preserves and wilderness areas surrounded by settled and degraded lands and waters (Foreman, *Rewilding North America*, 5).
3. Chawla, *Children and the Environment*, 22; Beyer et al., "Exposure to Neighborhood Green Space and Mental Health," 3455; Miller et al., "The Outcomes of Nature-Based Learning," 1118.
4. The concrete channel that contains the White River in downtown Indianapolis made an impression on me years before I ever visited Milwaukee. I remember seeing the dark-brown water flowing through the light-gray span of concrete and thinking it was the ugliest feature in the ugliest city I could ever imagine. My perception of the White River in 1990 remains, but my opinion of the city is changed by John Green, who offers an important context for recognizing the value of place and calling a place home (Green, *The Anthropocene Reviewed*, 163).
5. John Gurda, *Environmental Reckoning: Milwaukee Addresses Past Mistakes to Create More Promising Future*, Madison WI, accessed November 5, 2023, https://

widnr.widen.net/content/nmqycgubgv/pdf/GW_NRMagGreatLakesInsert_June 2020, 8.

6. Gurda, *Milwaukee*, 69.
7. Hayes, *Master Planners*, 54.
8. Hayes, *Master Planners*, 50.
9. "History," Milwaukee Metropolitan Sewerage District, accessed January 18, 2022, https://www.mmsd.com/about-us/history.
10. Paul G. Hayes, "Conservation and Environmentalism," Encyclopedia of Milwaukee 2016, accessed June 30, 2021, https://emke.uwm.edu/entry/conservation-and-environmentalism/.
11. Matthew Costello, "Milwaukee River," Encyclopedia of Milwaukee, accessed June 30, 2021, https://emke.uwm.edu/entry/milwaukee-river/; Kevin Mueller, "Revisiting the Riverwalk: 30 Years In," *Milwaukee Magazine*, February 28, 2018, https://www.milwaukeemag.com/revisiting-the-riverwalk/.
12. Norquist, *The Wealth of Cities*, 150.
13. Daniel, *The Milwaukee River Greenway*, 9.
14. Leinbach, *Urban Ecology*, 87.
15. "About Us," River Revitalization Foundation, accessed February 27, 2022, https://www.riverrevitalizationfoundation.org/about-us/; Daniel, *The Milwaukee River Greenway*, 34; River Revitalization Foundation, *The Riverway Plan*, 4.
16. Lisa Kaiser, "Extra Protections Created Milwaukee River Corridor," *Shepherd Express*, June 9, 2010, https://shepherdexpress.com/news/features/extra-protections-created-milwaukee-river-corridor/#/questions.
17. "Estabrook Dam Removal," Milwaukee Riverkeeper, accessed October 22, 2021, https://milwaukeeriverkeeper.org/dam-removal/estabrook-dam-removal/; *Milwaukee Greenway Master Plan*, Milwaukee Riverkeeper, June 2010, https://www.milwaukeeriverkeeper.org/wp-content/uploads/2015/04/Greenway_MasterPlan_Final_Report.pdf.
18. Daniel, *The Milwaukee River Greenway*, 37.
19. Daniel, *Urban Wilderness*, 107.
20. *Market Study, Engineering, and Land Use Plan for the Menomonee Valley 1998*, City of Milwaukee, Department of City Development, October 1998, https://static1.squarespace.com/static/5b1738a7f8370aa49cd05cf8/t/5b74646cb8a045311c66a3a9/1534354818129/mv+Land+Use+Plan+1998.pdf.
21. "Menomonee River Valley Neighborhood History," Menomonee Valley Partners, accessed December 4, 2021, https://www.thevalleymke.org/history.
22. "Analysis and Planning 2011 ASLA Professional Awards," American Society of Landscape Architects, accessed December 4, 2021, https://www.asla.org/2011awards/436.html.

23. In 2008 the UEC opened a new branch in Washington Park, where degraded woodland and a heavily polluted pond attracted the ambition of various neighborhood groups to invite the UEC. Washington Park is just over a mile north of the Menomonee Valley (Leinbach, *Urban Ecology*, 166).
24. Leinbach, *Urban Ecology*, 89.
25. Maher, *Milwaukee County's Oak Leaf Trail*, 2.
26. The plan took shape across multiple agencies and organizations over decades, but the *Regional Green Infrastructure Plan* (Milwaukee Metropolitan Sewerage District, June 2013, https://www.mmsd.com/static/MMSDGIP_Final.pdf) offers a high-level view of progress and ongoing efforts; "History," Milwaukee Metropolitan Sewerage District.
27. The program was established in 1997 and has grown into a central feature of the center's work in South Side neighborhoods of Milwaukee ("Environmental Health," Sixteenth Street Community Health Center, accessed January 16, 2022, https://sschc.org/support-programs/environmental-health/).
28. "Milwaukee Riverwalk," City as Living Laboratory, accessed January 16, 2022, https://www.cityaslivinglab.org/milwaukee-river-walk-1998-2001.
29. "Watermarks: An Atlas of Water for the City of Milwaukee," City as Living Laboratory, accessed January 16, 2022, https://www.cityaslivinglab.org/watermarks.
30. Mary Miss, in discussion with the author, June 16, 2018.

BIBLIOGRAPHY

Beyer, K., A. Kaltenbach, A. Szabo, S. Bogar, F. Nieto, and K. Malecki. "Exposure to Neighborhood Green Space and Mental Health: Evidence from the Survey of the Health of Wisconsin." *International Journal of Environmental Research on Public Health* 11, no. 3 (2014): 3453–72.

Chawla, Louise. *Children and the Environment*. New York: Oxford University Press, 2012.

Daniel, Eddee. *The Milwaukee River Greenway: A Wealth of Nature in the Heart of the City*. Milwaukee WI: River Revitalization Foundation, 2021.

———. *Urban Wilderness: Exploring a Metropolitan Watershed*. Chicago: Center for American Places, 2008.

Foreman, Dave. *Rewilding North America: A Vision for Conservation in the 21st Century*. Washington DC: Island Press, 2004.

Green, John. *The Anthropocene Reviewed: Essays on a Human-Centered Planet*. New York: Dutton, 2021.

Gurda, John. *Milwaukee: A City Built on Water*. Madison: Wisconsin Historical Society Press, 2018.

Hayes, Paul G. *Master Planners: Fifty Years of Regional Planning in Southeastern Wisconsin*. Milwaukee WI: Marquette University Press, 2010.

Leinbach, Ken. *Urban Ecology: A Natural Way to Transform Kids, Parks, Cities, and the World*. New York: Morgan James, 2017.

Leopold, Aldo. *A Sand County Almanac: And Sketches Here and There*. Oxford: Oxford University Press, 1949.

Maher, Jill Rothenbueler. *Milwaukee County's Oak Leaf Trail: A History*. Charleston SC: History Press, 2019.

Miller, N. C., S. Kumar, K. L. Pearce, and K. L. Baldock. "The Outcomes of Nature-Based Learning for Primary School Aged Children: A Systematic Review of Quantitative Research." *Environmental Education Research* 27, no. 8 (2021): 1115–40.

Norquist, John O. *The Wealth of Cities: Revitalizing the Centers of American Life*. Cambridge MA: Basic Books, 1998.

River Revitalization Foundation. *The Riverway Plan*. Milwaukee WI: River Revitalization Foundation, 1991.

Van Auken, Paul M. "Toward a Fusion of Two Lines of Thought: Creating Convergence between Aldo Leopold and Sociology through the Community Concept." *Socio-Ecological Practice Research* 2 (2020): 39–61.

7 The Lost River

The Revitalization of the Jordan River in the Salt Lake Valley

Marian L. Rice

The Jordan River, the main artery of Utah's Salt Lake Valley, possesses a rich history of stewardship, neglect, abuse, and revitalization. As with many urban rivers, it quickly transformed from a diverse and lush ecosystem to a channelized and polluted river that became lost for many residents. Likewise, some cities along the Jordan River struggled with ways to reimagine a river corridor lined with legacy industry and contamination, while other towns focused on keeping portions of the river natural while navigating the valley's looming development. Despite their varying priorities, communities in the Salt Lake Valley have converged, and the Jordan now serves as an example of how collective efforts and collaboration can revitalize a once-neglected waterbody.

Serving a booming population of nearly two million people, the Jordan River drains a 3,805-square-mile basin, which includes the Provo River, Spanish Fork River, and Utah Lake basins. The Jordan starts at the highly managed Utah Lake and flows fifty-one miles northward, ending its journey at the wetlands of the Great Salt Lake. Fourteen tributaries of the Jordan River flow from the Wasatch and Oquirrh Mountains. Like many urban rivers, the Jordan crosses multiple jurisdictions, including three counties and sixteen cities, all of which have varying uses and visions for the river. Furthermore, multiple regulatory authorities often have conflicting missions. For example, the Utah Department of Natural Resources division of Forestry, Fire, and State Lands (FFSL) considers the Jordan River a valuable ecosystem and manages the river to ensure ecological health, scenic attributes, and recreational opportunities. Yet, the three counties along the river have jurisdiction over flood control and prioritize the protection of human life and property. The flood-control jurisdiction includes controlling the alteration of natural floodplains, stream channels, and natural protective barriers. These conflicting missions add complexity to an already heavily managed river.[1]

The Jordan wasn't always so regulated. It was once a meandering river with a wide floodplain and an ecosystem that supported diverse aquatic and terrestrial life. Early inhabitants of the Salt Lake Valley relied on the river for food, water, and shelter, as it provided a source of livelihood for Native Americans and early settlers. As the population of the Salt Lake Valley increased in the late 1800s and early 1900s, so did the demands on the river. The Jordan River was straightened and channelized to meet the demands of industry and agriculture. Dams and canals were built to satisfy the increasing need for irrigation. By the early 1900s, the river became more of a polluted canal than a natural river corridor, and for most of the twentieth century, many residents avoided the river due to the polluted water and concerns about public safety. The 1970s environmental movement and ensuing legislation, coupled with the Provo River Parkway effort, initiated ways to reimagine and revitalize the river that continues today. While the Jordan drew upon efforts in other watersheds, it now serves as an example of how collective efforts can bring the community back to the river.

Rich History of an Urban Waterbody

Long ago the Jordan watershed was a much different place than what we see today. More than sixteen thousand years ago, ancient Lake Bonneville covered much of Utah. As Lake Bonneville receded, Utah Lake and the Great Salt Lake remained. The Jordan River emerged winding through unconsolidated Lake Bonneville sediments flowing northward from Utah Lake toward the Great Salt Lake.[2] Eventually, the Jordan River established a riparian habitat complete with oxbows, wetlands, ponds, and abundant wildlife.

The first known inhabitants to live along the Jordan River were members of the Archaic culture, approximately 8000 BCE to the beginning of the

10. Unidentified individuals swimming in the Jordan River, circa 1872. Courtesy of Utah State Historical Society.

common era. Subsequently, the Fremont culture, which had close ties to the environment of the Jordan River, occupied the valley. However, signs of the Fremont culture disappeared around 1300 CE. Later, the Shoshone, Ute, and Goshute Tribes inhabited the watershed. The tribes used the waterways as a major source of food with fish, plants, and wildlife that congregated around the waterbodies. The Ute relationship and love of the land tied their culture closely to the earth and its abundance. The Shoshone believed the streams, mountains, and plains stood forever, and the seasons walked around annually.[3]

Euro-American travelers passed through the Salt Lake Valley, but few stayed. However, in 1847 a contingent of pioneer members of the Church of Jesus Christ of Latter-day Saints entered the Salt Lake Valley and ushered in a new era for the Jordan. The settlers felt they were in the heart of a new holy land given the similarity of the Utah desert to Palestine. The Great Salt Lake corresponded to the Dead Sea, and thus the Jordan River flowing north from the freshwater Utah Lake acquired its name. Within days of arrival, the pioneers began to reshape the landscape.[4]

Later, the Jordan River became a site of industry and ore processing. In the 1860s silver ore was discovered in Bingham Canyon and sparked the beginnings of the smelting industry in the Salt Lake Valley. The majority of the smelters were located along the Jordan due to the water supply, which was essential for the smelting process, and access to the Wasatch Mountains and Bingham Canyon. The smelting industry conflicted with the agrarian culture of the original pioneers of the Salt Lake Valley. The "smelter smut" of acid rain and heavy metal-laden dust fell on the surrounding homes and fields, which burned and poisoned farmers' crops and stock. As a result of the impact on crops, there was an outcry from farmers and subsequent lawsuits. Nevertheless, the smelting industry along the river continued until 1958, leaving behind contaminated soils and groundwater.[5]

In the following decades, new homes and industry grew along the Jordan River. Since the settlement of the first pioneers, the communities of the Salt Lake Valley used the Jordan to dispose of their sewage and waste. Periodic flooding of the Jordan resulted in flood-control projects, including a diversion dam, dredging, and straightening of the river. Throughout the 1960s and into the 1970s, the Jordan continued to be used as a waste-disposal canal for industries, such as slaughterhouses, packing plants, mineral-reduction mills, and laundries.[6] Within a short period, the Jordan River transformed from a natural riverine system to an engineered and polluted channel, creating conflict with the people dependent on the river for substance. A river that people once flocked to became a lost river to be avoided and feared.

Rediscovery of a Lost River

The 1960s and 1970s marked a national environmental awakening. The establishment of the United States Environmental Protection Agency and passage of environmental legislation, such as the 1972 Clean Water Act, brought public attention to waterbodies. This renewed interest resulted in efforts to recognize the lost river and opportunities for revitalization. This initial effort was the result of a concerned citizenry coupled with the need of local government to accommodate flood control with a booming population in the Salt Lake Valley.

Driven by the need for flood control, in 1973 the Utah State Legislature created the Provo-Jordan River Parkway Authority to establish programs to enhance the natural quality of the river as well as develop park and recreational facilities, water conservation projects, and flood-control measures. Traditionally, authorities used a unilateral approach to flood control, which included channelization and minimal ecological benefits. However, the rapidly growing population in the Salt Lake Valley and the preservation of large, open space along the center of the valley pushed the Jordan River Parkway to embrace a new approach to old problems. As part of the parkway effort, two individual problems could be addressed with one solution: one, beautification and use of the Jordan for recreation and, two, control of the floodwaters. In addition to flood control and ecosystem enhancement, this approach could also address pollution, public health, and safety. The Army Corps of Engineers reached out to Salt Lake Valley citizens to create the Flood Control Advisory Committee to find methods to control the Jordan River without eliminating the multiple uses the public felt were important.[7]

The authority legislation included funding, although not to the extent needed to accomplish the lofty goals. The parkway efforts were slow and steady, but by 1986 approximately $18 million had been spent. Regardless of the initial successes, funding, especially private funding, was a constant challenge and a barrier to the continued projects along the Jordan River Parkway. As a result, in 2006 the Provo-Jordan River Parkway Authority was abolished and became known as the Riverway Enhancement Program, transferring all properties, rights, interests, powers, functions, duties, and obligations to the Division of Parks and Recreation.[8]

Other small efforts fizzled from lack of funding and other issues. However, the increased population growth and development in the Salt Lake Valley—coupled with the increased appreciation for the important environmental, social, and economic role of the river—compelled community leaders in 2007 to form the Jordan's largest planning initiative: the Blueprint Jordan River. Although Salt Lake County was the central player in this effort, mayors and staff from fifteen cities and three counties along the Jordan River contributed political, financial, and expertise support. The Blueprint steering committee consisted of members from the state government, local county and city government, private businesses, and nongovernmental organizations (NGOs). The Blueprint was the first comprehensive effort to develop a publicly supported vision for the future of the entire river and an implementation plan to transform the river corridor into a defining amenity for the Salt Lake Valley.[9]

The Blueprint included a significant public visioning effort to capture the community's collective imagination and build an appreciation for the important environmental, social, and economic roles of the Jordan River. Public involvement included surveys, workshops, and open houses. This effort received input from three thousand residents and technical experts. In December 2008 the Blueprint visioning process was complete. The Blueprint's vision incorporated the public's strong desire to preserve the remaining undeveloped areas along the Jordan and rehabilitate the river's natural functions. The Blueprint conveyed several "big ideas" and outlined a framework for how those could be implemented over the coming decades. Some of these ideas included building a contiguous trail between Utah Lake and the Great Salt Lake, removing obstacles to boat the river trail, preserving open spaces, creating neighborhood river centers, and establishing environmental education.[10]

Another outcome of the Blueprint Jordan River was the creation of the Jordan River Commission (JRC). The vision of the JRC was to facilitate the regional implementation of the Blueprint, to serve as a technical resource to local communities, and to provide a forum for coordination of planning, restoration, and responsible development along the Jordan River. Although it drew from the earlier example of the San Antonio River Authority, the main model of the JRC was the Utah Lake Commission. However, unlike the Utah Lake Commission, the JRC did not have the advantage of having local governments already vested in the project and was formed after the development of the Blueprint.[11]

In 2010 an Interlocal Cooperation Agreement formally established the JRC; however, it was an uphill process to develop the JRC into a successful and functioning commission. As the Jordan River had such a long history of failed efforts, some political entities and organizations were skeptical that the JRC would be successful. Furthermore, there was a perception that becoming a member of the JRC would limit what a city or county could do along the established boundaries of the river corridor. Therefore, there were only a few members initially, and it took time to gain trust. Also, the Blueprint did not define the role of the JRC. As a result, people and organizations created their own roles, resulting in differing thoughts and visions of the JRC's function. The environmental community saw the JRC as a way to hold cities accountable and pushed for the JRC to be an advocacy organization. Others thought the role was for land use, and some thought the JRC was a "sham" for commercial development. Cities along the corridor thought the role of the JRC was to complete the fragmented Jordan River Parkway Trail.[12]

It was clear the newly formed JRC had a big job ahead of it. One of the first goals was to keep the commission afloat and not fizzle as had previous efforts. The JRC had to persuade the cities that were initially apprehensive by creating non-threatening land-use guidance. Cities could define their own goals but still have the overarching vision of the 2008 Blueprint. In 2015 the JRC adopted its first strategic plan that included collaborative efforts by the members and communities surrounding the river. The same year the JRC members signed a new Interlocal Cooperative Agreement that modified the original 2010 agreement.[13]

The work of the JRC is directed by a governing board composed of representatives from member governments, appointed representatives of the state, and a group of NGO community partners. The JRC incorporates input and ideas from stakeholders, partners, and the public to help identify goals, priorities, and projects. The board guides complex decision-making and balances the multiple and sometimes conflicting expectations of projects around habitat, water quality, safety, and recreation. For example, balancing habitat and wildlife needs with public safety can present challenges. Vegetation provides shading and habitat for the river but can also be a recreational hazard for boats and kayaks floating on the river. Middle story habitat, such as shrubs and willows, can contribute to public safety concerns along the Parkway Trail. Therefore, projects and maintenance along the Jordan River corridor must balance these important, and at times conflicting, elements of recreation, public safety, and ecosystem integrity.

The JRC has had many successes since its inception, including gaining the trust of those entities that were initially apprehensive. For example, the first goal of the board was to complete the fragmented Parkway Trail along the Jordan River. The JRC facilitated funding efforts while working with local and state governments to find the means to complete all gaps in the trail. Through the collaborative work of municipalities, counties, and the JRC, gaps in the Parkway Trail were incrementally connected.

The final gap on the Parkway Trail was connected in November 2017, thus completing the longest paved urban trail in the western United States. Further, the JRC clarified its role, which is to serve as a capacity-building organization by leveraging funding opportunities, facilitating collaborative efforts, and leading the public's vision of the Blueprint. The JRC is currently comprised of citizens and government representatives from seventeen cities, two counties, six districts, and two state agencies. It does not have regulatory or maintenance authority for the Jordan River corridor, which lies with the federal, state, and local governments. Although the JRC is a governmental entity, it operates much like an NGO.[14]

With changes since the original 2008 Blueprint, the JRC decided to update the Blueprint to incorporate the accomplishments and look ahead to the next decade. As part of the update, the JRC issued public surveys to learn more about public awareness and priorities. Results from a public survey for the 2020 Blueprint Jordan River update demonstrate increased awareness and interest in the Jordan River corridor. For the 2020 update, eight thousand people participated in the online survey compared to three thousand participants in 2008. Respondents were younger and more diverse than in the earlier survey. The results also demonstrated an increased use of the river. Further, respondents felt there should be more investment in the Jordan River and the Parkway Trail.[15]

For the 2020 update, the JRC and Envision Utah made a concerted effort to engage diverse audiences. This is important as the Jordan River passes through communities with a high proportion of minority and low-income residents. For example, Salt Lake City is more racially and ethnically diverse than the state, with more than one-third (34.6 percent) of the population being non-white compared with 21.9 percent for the state. The city is divided into seven council districts. The Jordan River intersects districts 1 and 2, which are minority-majority at 64.6 percent and 69.2 percent respectively. These districts also house a significant portion of new immigrant groups, particularly refugee populations, further demonstrating the importance of the JRC's efforts to reach the communities most impacted by changes along the corridor.[16]

Public surveys demonstrate the Jordan as a place of relaxation, nature, and activity for the communities surrounding the river. The Hispanic community considers the Jordan an essential part of family recreation.[17] The river's importance is also reflected by the support from the state legislature, which continues to focus on improvements on the Jordan. For example, from 2018 to 2021, legislation has provided funding and support to the Jordan for recreational amendments such as boat launches and restoration.

Local governments work collaboratively with the JRC and other agencies to revitalize the Jordan River. One example is contributing to the Parkway Trail. Now that the Parkway Trail is complete, municipalities are adding trail sections on both sides of the river to increase accessibility. In 2018 Salt Lake County installed wayfinding signage along the Parkway, benefiting accessibility, use, and public health. The JRC is collaborating with Salt Lake County and local government to focus on the Jordan River water trail as a recreational, public health, and active transportation amenity. This collaborative effort includes the mitigation of navigational hazards, some of which date back to the mid-nineteenth century, and adding a series of new boat access points to increase accessibility.[18]

The increased public appreciation for the river's value continues to inspire renewed ecosystem restoration efforts. Salt Lake County and the local government continue to improve the ecosystem health and water quality of the Jordan River with watershed planning and ecosystem restoration projects. For example, Salt Lake County has implemented a variety of ecosystem restoration projects along the river that reconnect the river to the floodplain, provide habitat, and benefit water quality. Salt Lake County also hosts an annual watershed symposium, which is a comprehensive review of the current state of the Jordan River watershed and provides an opportunity to identify opportunities for collaboration.[19]

Salt Lake City actively works to improve the water quality and ecological health of the river. Salt Lake recently completed a stormwater and streambank ecosystem restoration project along the Jordan River that also incorporates education and improvements to the river trail. In concert with ecosystem restoration efforts, the city continues to implement measures to bring the community back to the river and change public perceptions. Lewis Kogan, director of Salt Lake City's Trails and Natural Lands division, feels most people make incorrect assumptions about the Jordan. He notes, "The Jordan River is already incredible in its current state and not as dirty and unsafe as people think. It is an issue of perception. Increasing accessibility, visibility, and getting more people to the river are all elements important to change perception and increase use."[20] Salt Lake City is doing just that by adding more trails and wayfinding signage connecting the river to the community and installing new boat launches. Salt Lake is also developing a self-serve kayak-sharing system that lets you rent a boat and paddle to a destination downriver. In 2019 the city received a National Park Service Rivers, Trails, and Conservation Assistance grant to help develop this paddle-share program.[21] The 2020 Blueprint Jordan River survey results and interviews demonstrate how these, and many other efforts, have increased the use and ownership of the river by the community.

NGOs bring a grassroots effort to revitalize the Jordan. The Jordan River Foundation was formed in 1979 to assist and cooperate with state and local government in the establishment and maintenance of the parkway as well as the development of recreation, water conservation, flood control, reclamation, and wildlife resources, and to actively generate broad public support and community interest. The foundation is one of many partners, including the JRC, local government, community partners, and other NGOs that jointly host the annual Get to the River Festival. This thirty-day festival brings together the public, community business partners, and local government to celebrate, protect, and restore the Jordan River Parkway. The event attracts thousands of volunteers and participants, bringing attention and inspiration to the Jordan River.[22]

Art is an impactful forum for the education and inspiration of the Jordan River. The Center for Documentary Expression and Art (CDEA) is a local NGO that uses oral history, photography, visual art, and music to help people better understand Utah's past and present and to discover the multifaceted connection to the nation and the world. In 2010 the CDEA proudly displayed the Tillman Crane "ReAwakened Beauty: The Past, Present, and Future of the Jordan River" exhibit with an accompanying residency program to heighten awareness of the history and potential of the Jordan River. In 2013 CDEA collaborated with JRC and Salt Lake County to develop the My Jordan River: Our River Our Future website (myjordanriver.org), which is an interactive map of the Jordan River corridor to further awareness and ownership of the river. The project also included the ReAwakened Beauty school residency program, which worked with underserved communities and schools along the Jordan River corridor.[23]

Conservation of public land plays an important role in the revitalization of the Jordan River. One such area is where the Fremont culture inhabited the Jordan River, with dwellings and artifacts, including stone tools and weapons, corn, and animal bones dating as far back as 800 to 480 BCE. Due to the importance of these discoveries, the state of Utah granted the FFSL title to an approximately 250-acre parcel of "critical land" adjacent to the Jordan River. This property is known as the Galena-Soónkahni Preserve, which means "many dwellings" in Shoshone. The land is preserved in perpetuity with a conservation easement held by Utah Open Lands, a local nonprofit land trust conservation association.[24]

Sometimes it is a single individual who realizes there is a problem and steps forward to do the right thing. The Jordan is fortunate to have such individuals who dedicate their life to bringing the community back to the river. Some of these individuals grew up along the river and learned about this urban treasure at a young age. They continue to enjoy the splendor the river offers, including fishing, swimming, kayaking, and urban farming. Some people may act as an ambassador and take anyone interested, including politicians, decision-makers, and the public, out on the Jordan in a kayak to show them the positive and negative aspects of the river, always pointing out opportunities and solutions. Others facilitate trash pickups and weed pulls. Some may have lifelong experiences and knowledge to share, which benefits revitalization efforts. These individuals have affected the positive change in public perception and water policy along the Jordan River. One such advocate is Elliott Mott, an outdoor enthusiast and activist for the Jordan River. To introduce people to the splendor of the Jordan, Mott hosts group kayak float trips and bike rides. In 2018 Mott published *Jordan River Water Trail*, a guidebook that provides detailed instructions for locating sparse river access points. When reading through the book, it is apparent Mott is passionate about the Jordan and wants to share his vision with anyone who will listen.[25]

Another dedicated advocate for the Jordan is councilmember Tish Buroker, whose stated reason for going into politics is her passion for the river. Buroker said when she was a child, she never went down to the river and remembered it as a place where you throw away trash. In 1994 she bought a house along the banks, and after moving in, Salt Lake County and the city of Riverton restored a wetland along the river, which she described as a beautiful asset. However, Buroker moved away for a period and, when she returned to her home, was stunned by how degraded the wetland looked. This propelled Buroker to get involved and started attending city council meetings to advocate for the Jordan. She ultimately decided to run for a seat on the Riverton City Council. As a council member, she advocates for the river and sits on the board of the JRC. According to Buroker, the many efforts, especially the completion of the Jordan River Parkway Trail, have been a "game-changer" and opened people's minds.[26] Further, she finds the negative stigma of the Jordan River is generational and that younger people and those who did not grow up in Utah have a more positive perception of the river. Also, many who live along the river love the Jordan and appreciate having such a piece of nature in their backyard.

The Jordan River has transitioned from a robust natural ecosystem used by Native Americans to a polluted and lost waterway avoided by many to, finally, an appreciated urban river parkway. The effort to reimagine the Jordan River started in the 1970s with the Provo-Jordan River Parkway Authority, as a way to manage flooding while providing a recreational amenity. Later, in the mid-2000s, the Blueprint Jordan River vision effort resulted in the development of the Jordan River Commission and increased awareness of the river. The collaborative efforts of the JRC, local government, the state, NGOs, and the community are all essential to the revitalization of the river.

Nevertheless, the years of neglect and abuse from the channelization and dredging in the late 1800s to the industry and smelting in the early 1900s were not kind to the river. The Jordan River is still contaminated and far from its natural ecosystem. However, it has come a long way from years of being a lost river and a place of refuse. There is an increased momentum of ecosystem restoration and water-quality projects as well as many new opportunities for people to enjoy the river, such as the proposed paddle share, new trails, and crowdsourcing like the My Jordan River: Our River Our Future website. By increasing access to the Jordan, such as with trails, boat launches, and information, the river serves as an alternative means of transportation with people using the river to commute as well as for recreation.[27]

The Jordan River continues to shake its negative stigma, while positive perception continues to grow. The recent 2020 Blueprint Jordan River surveys demonstrate more people are enjoying the Jordan River Parkway and feel it is a positive amenity. The communities that live along the river and people from all over enjoy taking advantage of the Jordan River Parkway Trail.[28] According to Laura Hanson, the former executive director of the Jordan River Commission, "The Jordan River is equitable; it is something for everyone."[29] With continued efforts working in concert, the once-lost Jordan River will continue to be an urban treasure for current and future generations.

NOTES

1. Salt Lake County, *Water Quality Stewardship Plan*; Soren Simonsen (director of the JRC), in discussion with the author, December 2019.
2. *Blueprint River Jordan*, Salt Lake City UT, 2008, http://jordanrivercommission.com/wp-content/uploads/BlueprintJordanRiver_LowRes3.pdf.
3. Cuch, *History of Utah's American Indians*, 7; Duncan, "The Northern Utes of Utah," 67–224; Parry, "The Northwestern Shoshone," 26; Rice, "Energy Water Nexus," 25.
4. Alexander, "Irrigating the Mormon Heartland," 174; Bancroft, *History of Utah*, 258–59; Alexander, "Interdependence and Change," 292; Meade and Teele, *Report of Irrigation Investigations in Utah*, 42–47.
5. Hughes, "The Development of the Smelting Industry," 48; Church, "Smoke Farming," 197–98; Lamborn and Peterson, "The Substance of the Land," 308–25; "Midvale Slag Midvale UT," U.S. Environmental Protection Agency, accessed November 23, 2019, https://cumulis.epa.gov/supercpad/cursites/csitinfo.cfm?id=0800641.
6. K. B. Hallaran, "Jordan River," Utah Education Network, accessed December 5, 2019, https://www.uen.org/utah_history_encyclopedia/j/JORDAN_RIVER.shtml.
7. Urban Technology Associates, *Jordan River Parkway*, 15; Hart Wixom, "Jordan River Need: A Master Plan Plus Money," *Deseret News*, December 28, 1971; Jerry Johnston, "Jordan Parkway: Two Rivers in Search of a New Identity," *Deseret News*, January 22, 1986.
8. Johnston, "Jordan Parkway"; Provo-Jordan River Parkway Authority transferred to Division of Parks and Recreation, UCA 79-4-801, Salt Lake City UT, 2019.
9. National Audubon Society, *The Jordan River Natural Conservation Corridor*, 32; "The River," Jordan River Commission, accessed November 23, 2019, https://jordanrivercommission.com/the-river/.
10. *Blueprint River Jordan*, 2008.

11. "About the Utah Lake Commission," Utah Lake Commission, accessed November 23, 2019, http://utahlakecommission.org/about-us/; *Blueprint River Jordan*, 2008.
12. "About the Jordan River Commission," Jordan River Commission, accessed November 23, 2019, http://jordanrivercommission.com/about/; Laura Hanson (former executive director of the JRC), in discussion with the author, November 2019; Rice and McCool, "Collaboration and the Criteria for Success," 17.
13. Hanson, discussion; "About the Jordan River Commission," Jordan River Commission; "The River," Jordan River Commission.
14. "Jordan River Parkway," Salt Lake County, accessed November 23, 2019, https://slco.org/parks-recreation/parks-trails/jordan-river-parkway/; Simonsen, discussion.
15. "Blueprint Jordan River Refresh," Envision Utah, accessed November 21, 2020, https://envisionutah.org/blueprint-jordan-river-refresh.
16. Salt Lake County, *Integrated Watershed Plan*, 1; *Salt Lake City Data Book*, Kem C. Gardner Policy Institute, Salt Lake City UT, 2020, https://gardner.utah.edu/wp-content/uploads/SLC-Data-Book-2020forWeb.pdf?x71849; Hubbard, "The Role of Knowledge in Water Resource Management," 592.
17. Carothers, "Justice and the River," 143.
18. Simonsen, discussion; "Water Trail," Jordan River Commission, accessed November 23, 2019, http://jordanrivercommission.com/water-trail/; "Jordan River Parkway," Salt Lake County, accessed November 23, 2019, https://slco.org/parks-recreation/parks-trails/jordan-river-parkway.
19. Salt Lake County, *Integrated Watershed Plan*, 1; "Watershed Symposium," Salt Lake County, accessed November 23, 2019, https://www.slco.org/watershed/watershed-symposium/.
20. Lewis Kogan (Salt Lake City Public Lands director), in discussion with the author, December 2019.
21. Kogan, discussion.
22. "The Jordan River Foundation," Jordan River Foundation, accessed November 23, 2019, https://jrf-utah.org/; "Get to the River Festival," Get to the River Festival, accessed December 5, 2019, https://gettotheriver.org/.
23. "ReAwakened Beauty: The Past, Present, and Future of the Jordan River," Center for Documentary Expression and Art, accessed November 23, 2019, https://cdeautah.org/exhibits-that-teach/; "A Walk along the Jordan," Tillman Crane, accessed November 23, 2019, https://www.tillmancrane.com/portfolio_jordan; "My Jordan River," Jordan River Commission, accessed November 23, 2019, https://myjordanriver.org/.
24. Jason Lee, "Draper Monument Pays Tribute to Utah's Native American Heritage," *Deseret News*, March 27, 2015; Utah Forestry Fire and State Lands, *Jordan River Comprehensive Management Plan*, 5.
25. McCool, *River Republic*, 12; Mott, *Unlocking the Mystery*, 3.

26. Trish Buroker (Riverton City Council member), in discussion with the author, March 2020.
27. Utah Division of Water Quality, *Jordan River Total Maximum Daily Load*, 68; Utah Division of Water Quality, *Utah's Final 2016 Integrated Report*, 18; "My Jordan River," Jordan River Commission.
28. "Blueprint Jordan River Refresh," Envision Utah; Carothers, "Justice and the River," 170.
29. Hanson, discussion.

BIBLIOGRAPHY

Alexander, Thomas G. "Interdependence and Change: Mutual Irrigation Companies in Utah's Wasatch Oasis in an Age of Modernization, 1870–1930." *Utah Historical Quarterly* 71, no. 4 (2003): 292–314.

———. "Irrigating the Mormon Heartland: The Operation of the Irrigation Companies in Wasatch Oasis Communities, 1847–1880." *Agricultural History* 76, no. 2 (2002): 172–87.

Bancroft, Hubert Howe. *History of Utah: 1540–1886*. New York: Arno Press, 1889.

Carothers, Taya L. "Justice and the River: Community Connections to an Impaired Urban River in Salt Lake City." PhD diss., Utah State University, 2018.

Church, Michael A. "Smoke Farming: Smelting and Agricultural Reform in Utah, 1900–1945." *Utah Historical Quarterly* 72, no. 3 (2004): 196–218.

Crampton, Gregory. "Indian Country." *Utah Historical Quarterly* 39, no. 2 (1971): 90–94.

Cuch, Forrest S. *History of Utah's American Indians*. Logan: Utah State University Press, 2003.

Duncan, Clifford. "The Northern Utes of Utah." In *History of Utah's American Indians*, edited by Forrest S. Cuch, 167–224. Logan: Utah State University Press, 2003.

Hedges, A. H. "News from Salt Lake, 1847–1846." *Utah Historical Quarterly* 84, no. 3 (2016): 217–36.

Hubbard, Monica L. "The Role of Knowledge in Water Resource Management: An Assessment of the Oregon General Public." *Social Science Journal* 60, no. 3 (May 2, 2020). https://doi.org/10.1080/03623319.2020.1782635.

Hughes, Charles E. "The Development of the Smelting Industry in the Central Salt Lake Valley Communities of Midvale, Murray, and Sandy Prior to 1900." MA Thesis, Brigham Young University, 1990.

Lamborn, John E., and Charles S. Peterson. "The Substance of the Land: Agriculture vs. Industry in the Smelter Cases of 1904 and 1906." *Utah Historical Quarterly* 53, no. 4 (Fall 1985): 308–25.

McCool, Daniel. *River Republic: The Fall and Rise of America's Rivers*. New York: Columbia University Press, 2012.

Meade, Ellwood, and Ray Palmer Teele. *Report of Irrigation Investigations in Utah.* Washington DC: Government Printing Office, 1904.
Mott, Elliott R. *Unlocking the Mystery of the Jordan River Water Trail and Bike Path.* Salt Lake City UT: Roosevelt and Torrey, 2018.
National Audubon Society. *The Jordan River Natural Conservation Corridor Report.* Salt Lake City UT: U.S. Fish and Wildlife Service, 2000.
Parry, Mae. "The Northwestern Shoshone." In *History of Utah's American Indians*, edited by Forrest S. Cuch, 25–72. Logan: Utah State University Press, 2003.
Powell, Allan Kent. *Utah History Encyclopedia.* Salt Lake City: University of Utah Press, 1994.
Rice, Marian L. "Energy Water Nexus: An Analysis [of] How the Energy Policy Act of 2005 and Energy Extraction [Have] Affected the Uinta Basin and the People Who Live There." PhD diss., University of Utah, 2021.
Rice, Marian L., and Daniel McCool. "Collaboration and the Criteria for Success: A Case Study and a Proposed Framework for Analysis." *Administration & Society* 54 (September 2021). https://doi.org/10.1177/00953997211042564.
Salt Lake County. *Integrated Watershed Plan.* Salt Lake City UT: Salt Lake County Flood Control, 2016.
———. *Salt Lake County Water Quality Stewardship Plan.* Salt Lake City UT: Salt Lake County Flood Control, 2009.
Urban Technology Associates. *Jordan River Parkway: An Alternative.* Salt Lake City UT: Board of County Commissioners, 1971.
Utah Division of Water Quality. *Jordan River Total Maximum Daily Load Water Quality Study Phase 1.* Salt Lake City: Utah Department of Environmental Quality, 2013.
———. *Utah's Final 2016 Integrated Report.* Salt Lake City: Utah Department of Environmental Quality, 2016.
Utah Forestry Fire and State Lands. *Jordan River Comprehensive Management Plan.* Salt Lake City: Utah Department of Natural Resources, 2017.

Part 3 CONVERGENCE

Old Rivers, New Plumbing

Accepting the premise of the Anthropocene, part 3 explores how the merging of human and natural systems is manifested on the landscape. In his highly regarded 1995 book, *The Organic Machine: The Remaking of the Columbia River*, historian Richard White argues that the entire Columbia River system is controlled and managed by humans yet remains a wild entity; the river is both natural and technological—essentially a cyborg. In addition, White maintains that the Columbia River exists as a network of social relations, both between different groups of people and between humans and nature. Whether explicit or not, the authors in part 3 apply the "Organic Machine" as a foundational concept in rewilding urban rivers. The authors elaborate on White's ideas and explore the hubris of technocratic approaches to alleviating the human impacts on urban rivers, yet they each draw different conclusions.[1]

In "Engineering Nature," Brian Murphy discusses how Denver's South Platte River suffers from "urban stream syndrome" and how the engineering efforts of designing with nature rather than against it has led to substantial improvements in river health. He highlights the seismic shift within the Army Corps of Engineers from an institution traditionally focused on the control of nature through dams and channelization to one that has embraced ecological restoration.

In the case of the South Platte, as with many urban rivers, restoration to its former self is not an option. Thus, Murphy argues that any path toward improvement will entail some form of engineering and that degraded natural systems require enhancement to fulfill human desires. Like White, Murphy recognizes that rivers are "complex socio-ecological systems," but also that engineers operate within this system. As our values shift, in this case from dominating to working with nature, so do the tasks we ask of engineers. Designing with nature means balancing competing interests, such as flood control, economic development, recreation, and ecosystem health, through collaborations and partnerships. For Murphy, nature itself becomes one of the many collaborators in urban rewilding.

Accepting the limitations of full-blown restoration, Murphy introduces the concept of ecological lift as a means of improving ecological functionality to move urban rivers from degraded stream syndrome and toward a healthier trajectory. On the Deschutes River in Oregon, ecological lift is quite literal as Kirsten Rudestam, in "Cyborg Salmon and Plumbed Basins," highlights the salmon elevator at the Pelton Dam to illustrate how this iconic fish has become part of the Organic Machine, or what Rudestam terms "waterworld"—a highly complex system of plumbing and water transfers.

In waterworld, the Deschutes becomes a simulation of nature hidden behind the façade of a wild river. As one water manager notes, "Every drop of water in this basin is managed." A sociologist, Rudestam seeks to unpack this inherent conflict between a managed watershed and our attachment to wild nature. Drawing upon the work of Donna Haraway, Rudestam argues that salmon in the Deschutes are cyborgs—such that the line dividing living beings and technology is sufficiently muddled so that the distinction is difficult to ascertain. Likewise, with urban rivers the boundaries between human and natural have dissolved and the two are now interwoven. Ultimately, our conceptions of both nature and technology are cultural products and neither are free from human social relations.

Perhaps the original cyborg river dates to 1900 when Chicago succeeded in reversing the flow of its river and connecting the St. Lawrence watershed to the Mississippi. As Shawn Bailey in "Neon River" tells us, the Chicago River is America's most humanized riverscape and an "invasive species superhighway." With seventy-eight miles of canals and 70 percent of its flow treated wastewater, the Chicago is truly a "river of the Anthropocene." Nonetheless, dedicated efforts for ecological lift on the Chicago River have led to a metropolitan rewilding of beavers, otters, coyotes, and fifty-eight different fish species in and around the river. To forestall the worst impacts from invasive species moving from the Great Lakes to the Mississippi drainage and vice versa, the most ambitions restoration plan calls for re-reversal of the Chicago River so that it would flow into Lake Michigan once again. This $15 billion proposal from the Army Corps of Engineers would take a quarter century to complete, but, as Bailey points out, given the anticipated and unforeseen impacts of climate change, this might not even work.

Rather than replicate the hubristic and technological approaches of the past, Bailey suggests reimagining what the river is for. For Bailey, a river is more than a conveyance to move water; it has an intrinsic right to exist, and, therefore, the way to approach rewilding in the Anthropocene is to provide legal personhood to our rivers. Granting rights to rivers is certainly a Chicago-sized legal lift but one that is slowly gaining acceptance across the globe. In 2017 New Zealand granted legal personhood to the Whanganui River, and two years later Bangladesh's high court granted rights to all its rivers. Meanwhile, in the United States, Toledo, Ohio, passed a bill of rights for Lake Erie, and the Yurok Tribe of northern California has granted tribal rights to the Klamath River.

All of this could be predicated on the prescient dissent from associate justice of the Supreme Court William O. Douglas in *Sierra Club v. Morton*. In 1972 Douglas argued that natural objects should have legal standing. Specifically calling out rivers, he wrote, "The river, for example, is the living symbol of all the life it sustains or nourishes—fish, aquatic insects, water ouzels, otter, fisher, deer, elk, bear, and all other animals, including man, who are dependent on it or who enjoy it for its sight, its sound, or its life. The river as plaintiff speaks for the ecological unit of life that is part of it."[2] Rewilding in the Anthropocene will require novel ways of thinking, and rethinking our legal system is a place to start.

NOTES

1. See White, *The Organic Machine*.
2. Sierra Club v. Morton, 405 U.S. 727, 92 S. Ct. 1361 (1972).

BIBLIOGRAPHY

White, Richard. *The Organic Machine: The Remaking of the Columbia River*. New York: Hill and Wang, 1995.

8 Engineering Nature

Denver and the South Platte River

Brian M. Murphy

Like many of America's urban rivers, the South Platte has been a working river for most of its post-settlement history. But before it started working for the people of Denver, Colorado, the South Platte River was quite different. The unconfined river frequently overflowed its banks and meandered across a broad floodplain. Following the discovery of gold in 1858, the South Platte River began to support agricultural, municipal, and industrial uses. At first, the South Platte through Denver served as a water transportation route for supplies, but industrial and residential development quickly encroached into the floodplain.[1] By the end of the nineteenth century, the primary uses of the river were for sewage disposal and efficient conveyance of water. The demand for water uses continued to expand with growth in population and land development in the Denver metropolitan area throughout the twentieth century. Uncannily prescient in capturing the ensuing trajectory of the river, a 1927 article titled "The Taming of the Treacherous Platte" noted, "The charm and beauty of the scenery along the [South] Platte River has a unique appeal—in its utter difference from any other part of Denver. It is believed that once the Platte has been subdued, there will be no more picturesque section of the city than along its banks."[2] The South Platte was, and still is, essential for wastewater effluent discharge and floodwater conveyance, wildlife habitat, and recreation.

Multiple landscapes, climate zones, and varying geologic formations shape the South Platte, as it flows four hundred miles from the eastern flank of the Rocky Mountains to its confluence with the North Platte River in Nebraska. The river bisects Denver, the center of a rapidly growing metropolitan area with a semiarid climate, approximately a hundred miles into its journey. As an urban river, it is also a product of its surrounding community. Denver's origins are tied to the banks of the South Platte, and subsequent smelting of gold and silver in the 1870s and 1880s initiated the long decline in water quality. Periodic flooding along the South Platte and its major tributaries marked Denver's early years, and as the city grew, the river became a sewer full of human waste and a conduit for conveying floods and irrigation water. By the early twentieth century, Denver's leaders began to consider engineering the iconic river for flood control and water supply. From 1898 to 1932, the Denver Union Water Company built dams and reservoirs, including Eleven Mile Canyon and Cheesman in the mountains above Denver on the South Platte River, to ensure an adequate water supply for the growing city.[3] However, floods continued to plague the city. Enter the U.S. Army Corps of Engineers.

The Flood Control Act of 1917 authorized the Corps to undertake, on a watershed basis, studies and surveys for flood-control improvements and to provide information regarding the relationship of flood control to navigation, waterpower, and other uses.[4] The Flood Control Act of 1936 further expanded that power and set up a two-pronged approach to reducing flood damages. The Department of Agriculture would develop plans to reduce runoff and retain more rainfall where it fell, while the Corps would develop engineering plans for downstream projects.[5] The Corps' first downstream flood-control project in the Denver metropolitan area was Cherry Creek Dam and Reservoir. Floods often turned Cherry Creek from an "inconsequential dry channel into an enemy for the city to defeat."[6] In 1948 Cherry Creek Dam became the first of the three dams the Corps would build to lower the risks to the Denver region from catastrophic South Platte River floodwaters that plagued the area for more than one hundred years.[7]

The worst flooding in Denver's recorded history occurred in 1965, when the South Platte River overflowed its banks and caused $500 million of damage, $300 million in the Denver area alone. The 1965 flood triggered the release of long-delayed flood-control projects, such as dams, reservoirs, and levees. The Army Corps and local agencies, like the Mile High Flood District (formerly the Urban Drainage and Flood Control District), began the process of "repairing" the river and its tributaries, which were channelized, straightened, and disconnected from their floodplains to reduce flood hazards to surrounding neighborhoods and businesses. Such efforts provided the driving force for ambitious urban-renewal plans and the beginning of recreation in and along an abused and neglected river, further reengineering the South Platte River. While critical to public safety, such infrastructure became a straitjacket on the river's ecological processes. Diversion of flow from channels, storage of water in reservoirs, and increased use of groundwater affected the distribution and timing of stream flows and the transport of fluvial sediments. By the second half of the twentieth century, the Denver reach of the South Platte was an industrial sewer. In addition, gravel miners had ravaged the river channel causing adverse geomorphic changes along the river corridor. All of this propelled the South Platte toward urban-stream syndrome—degraded physically, chemically, and biologically.[8]

Denver's growth continued to encroach into historic stream and floodplain habitats along the South Platte River and most of its major tributaries. By 1960 metropolitan Denver had grown to 164,000 acres with a population of 800,000. By 1980 the population surpassed 1,600,000. Today, metro Denver's population is nearing three million. Such rapid growth required the development of water-transfer projects to supply freshwater for domestic use. The Corps and the Bureau of Reclamation constructed dams and reservoirs in nearby hills and mountains, along with canals and tunnels to transfer water from the Western Slope of Colorado to the Eastern Slope (termed trans basin diversions). In 1975 the Corps built Chatfield Dam on the mainstem of the South Platte River to provide flood control as a reaction to the 1965 flood. The Corps constructed another flood-control dam in metropolitan Denver on another major tributary to the South Platte River—Bear Creek. Bear Creek Dam was the last of the three flood-control facilities to reduce the flooding risk in the Denver area.

The Corps began "taming the treacherous Platte" in earnest. The Corps had played an important role in shaping water resources systems in Colorado since Congress first tasked it in 1931 to prevent flooding along Cherry Creek. Since then, the Corps modified the South Platte among other streams in Colorado for flood-control purposes. As the Corps advanced its mission of delivering vital public and military engineering services and reducing risks from natural disasters, the South Platte felt the effects of such flood-control projects, with resulting changes in channel geometry and removal of riparian vegetation. The physical modifications and addition of treated wastewater effluent caused damages to natural resources that were not recognized until the mid-1990s when a new approach of river management started. The increasing awareness of the river as a valuable natural resource curated a new vision for the South Platte, ushering in a focus on insuring flow and water-quality levels needed to support recreational and environmental needs.[9]

The story of the South Platte River now includes engineering practices that are nature-based, rooted in understanding of ecological processes, and driven by community values. Those practices include constructing a standing wave for surfing, improving water quality in central Denver so that the river is fishable and swimmable, and restoring open space and instream aquatic habitat in the river north of Denver. This is also a story about a river that continues to be the lifeblood of the communities along its banks. It is a much different river than the one pioneers encountered during the 1859 gold rush. The evolution of river-engineering practices along the South Platte River has transformed it from an ephemeral "inch-deep, mile-wide" stream, to a flood-passing conduit, to an engineered urban ecosystem.

Postflood Solutions and New Problems

Progressive disruption of urban areas in the Denver metropolitan area by the South Platte River and its tributaries, alongside limited awareness of the inherent dangers of living on floodplains and overconfidence in these actions, brought about a major shift in management actions in the mid- to late twentieth century to rectify these problems.

The South Platte was an intermittent river prior to water development and the construction of the flood-control facilities on Bear Creek, Cherry Creek, and on the South Platte River at Chatfield. Historically, flows along the South Platte River were variable, rising spring runoff pulses from snowmelt in the Rocky Mountains and lowering to almost dry conditions during other times of the year. Prior to the flood-control era (1940–80), it carried little water during the summer other than following severe rainfall events, and it dried up completely in some years. Dams and their reservoirs, however, essentially reversed this pattern, lowering the base flows during spring, while increasing the summer flows, converting the river from intermittent to perennial. In addition to the dams, urbanization of the floodplain and conversion of prairie vegetation to residential and commercial developments also altered runoff patterns.[10] The conversion of the South Platte to a perennial river also influenced the natural sediment patterns. Previously, sediment movement from upstream to downstream maintained an alluvial conveyor belt that sustained instream aquatic habitat such as riffles, point bars, and channel braiding, but eliminating or reducing the frequency of episodic floods limited the transport of gravels and cobble from the headwaters to eastern Plains.

While the changes to the flow and sediment regimes certainly decreased the frequency of flooding in Denver (although major flooding can still occur in the urban areas), it significantly altered the South Platte River's aquatic and riparian ecosystems. The channel has become less complex structurally, decreasing fish habitat, while the lack of vegetation in the riparian corridor reduced shading and vegetative detritus leading to higher water temperatures. As flood peaks diminished, spring runoff events were no longer capable of transporting sediment and scouring vegetation from sandbars, progressively narrowing the channel.[11] Overbank flows became less common due to channel narrowing, so vegetation growth increased within the channel, further reinforcing the historical river character from braided to single thread.

Certainly, many of these changes to the river adversely affected the ecosystem. On the other hand, the South Platte is now a perennial river, and the consistency of low flows now supports a more diverse fish population (including rainbow trout). While the flow and sediment regimes are different—and, some argue, forever degraded—urban rivers are often the last natural areas in the built environment. The changes have also led to improved public safety, property protection, and ecosystem services as well as increased access to nature for underrepresented communities. The control of nature through flood mitigation created a novel hydrology and geomorphology, such that a return to past conditions is now impossible, and while there are multiple pathways forward for the South Platte, all require some form of engineering nature.

Channelization

Before three million people surrounded its banks, the South Platte was wider and shallower, more sinuous, and subject to major yearly flooding that formed braided channels.[12] In its pristine state, the river adjusted its width, depth, and sinuosity in response to the amount of incoming water and sediment. Alternating cycles of drought and wet years influenced the river's form.[13]

Changes in the physical condition of the riverscape indicated ongoing changes in streamflow patterns. As the river adjusted to new flow and sediment regimes, vertical banks sloughed into the water taking trees, willows, and other riparian plants into the channel. Downstream of Chatfield Reservoir, the river transitioned from rural to urban as it approached Denver.[14] As urbanization spread up and down the watershed, the river began to degrade and erode. Pools, riffles, overhanging banks, and other instream habitat features were changed or permanently lost. The changes in flow and sediment regime due to urbanization, dams, channelization, and in-channel mining disturbed the dynamic conditions associated with natural streams. The Corps and local agencies implemented a variety of techniques to manage this instability and erosion along the South Platte and its tributaries. Most of these techniques were structural and designed to arrest channel incision and bank erosion. Without channelizing the river and adding structures, like grade control and bank armoring, natural river processes were overwhelmed with incision, over-steepened banks, and loss of riparian vegetation. The river was in a state of downward evolution and needed help to restore its natural processes.

A primary example of channelization along the South Platte occurred in Littleton, ten miles south of downtown Denver. Seeking to accommodate larger flood flows, the Corps lowered the riverbed, widened and straightened the existing channel, stabilized the banks with concrete rubble and boulders, and disconnected the channel from its floodplain, essentially locking the river in place. The Corps also installed grade-control structures to prevent incision and protect infrastructure along and under the riverbed.[15] As natural stabilization and reestablishment of vegetation can take nearly a decade, the Corps implemented bank stabilization projects. However, channelization and construction of grade-control structures can

11. The South Platte River as it flows through downtown Denver, Colorado. Like all rivers, the South Platte is a product of its landscape, and as an urban river, it is a product of its surrounding community. Courtesy of Mile High Flood District.

increase stream velocity and reduce instream aquatic habitat. The channel becomes extremely uniform with little to no instream aquatic habitat or bank cover (e.g., trees, shrubs). Moreover, local agencies must maintain the channel, removing woody vegetation to preserve the flood-control capacity of the channel.

One of the prime reasons to channelize and stabilize rivers and creeks in urban areas is to protect the massive investment in infrastructure and private property. The Mile High Flood District (MHFD) protects a vast network of infrastructure within the South Platte River Corridor, which includes floodplain hazard mitigation works (levees, dams, reservoirs), streamside infrastructure (roads, trails, pipelines, streambank protection), and stream-crossing infrastructure (bridges, culverts, pipelines, grade-control structures, surface-water diversion structures) (see fig. 11). This infrastructure provides vital social services, but it is often detrimental to stream ecosystems, creating hazardous conditions (e.g., levee breaching) and being high maintenance. When much of this infrastructure was built in the mid- to late twentieth century, engineers prioritized stream hydrology and hydraulics and designed hard engineering structures and protection measures with little consideration of fluvial processes and stream ecology. This approach, coupled with aged infrastructure, resulted in high maintenance costs and adverse ecological impacts.

Natural Resource Damages

The combination of flood-control infrastructure, urbanization, and agricultural water demands permanently interrupted the natural flow regime and prairie ecosystems. When the stream flow was intermittent, woody vegetation was sparse along the river corridor, and the region was dominated by shortgrass and midgrass prairie species with cottonwood groves alternating with prairie on the floodplain. But irrigation practices raised the water table, and a continuous streamflow caused an increase in riverbank and floodplain vegetation. The river metamorphized from braided channel to a single-thread channel due to the changes in the water and sediment inputs.[16] As a result, much of the riparian and aquatic habitat along the South Platte River in the Denver area transitioned from a wide and shallow channel with gravel and sandbars with a few willows and cottonwoods, to a narrow and deep channel with limited gravel and sandbars with narrow bands of riparian vegetation.[17]

Channelization, bank stabilization, construction of grade-control structures, and modifications to the surrounding floodplain all contributed to altered flow and sediment regimes with degraded water quality and aquatic habitat. The water quality degradation compromised the ecological integrity and may have eliminated many of the native fish species in the South Platte River. The introduction of non-native fish into Colorado early in the twentieth century to support recreational fishing also changed the species composition of the ecosystem.[18] Colorado Parks and Wildlife (formerly Colorado Division of Wildlife) has continued this practice over the last forty years, along with both historical and more recent introductions of predator-type fish like the common carp. This practice has likely negatively impacted the native fish community. The combination of introduced aquatic species, degraded water quality, and severe alteration of the physical habitat has created a novel river ecosystem that is now outside the range of historical variability and beyond natural recovery.

The South Platte required an innovative engineering approach to meet this environmental challenge.

A New Era of River Management

A catastrophic event, like the 1965 flood, can cause rapid and significant changes to a river like the South Platte, as it ripples through the surrounding community, destroying homes, infrastructure, and habitat. A major flood can also galvanize the community. One of the most important developments to come out of the '65 flood was the creation of the Platte River Greenway, an effort led by Joe Shoemaker, a conservation-minded, former Republican state senator in the Teddy Roosevelt tradition. Shoemaker established the Greenway Foundation in 1974, which played a key leadership role in conceptualizing the South Platte as an amenity to reclaim, nurture, and protect. From 1975 to 1983, the Platte River Greenway served as a national model for urban rivers, changing the paradigm from controlling the river to engaging the river.[19]

Visionaries of the South Platte River, like Shoemaker and former Denver mayor Wellington Webb, instituted a restoration plan for the South Platte River. In 1995 Webb endorsed a set of recommendations for enhancing Denver's South Platte River Corridor and created the South Platte River Commission. "A Vision for the South Platte River" offered improvements to multiple uses of the river: recreation, boating, fishing, open space, wildlife habitat, education, youth employment, community awareness, and flood control.[20] Over the past twenty years, communities, state and local governments, federal agencies, and nonprofits like the Greenway Foundation invested significantly in "engineering nature" along the South Platte River to implement the vision.

The restoration of the South Platte, city planners wrote, would not only bring beauty to a long-ignored part of the city but also turn it into a "significant natural amenity . . . [with] the potential to become the cohesive force which draws the region's citizens together."[21] Waiting for the return of natural processes along urban rivers like the South Platte is often not an option due to public safety concerns with high-energy flood hazards. Jump-starting the restoration process through "engineering-with-nature" options provided opportunities to integrate natural and engineered systems efficiently and effectively.[22]

In the early 1990s, the Corps expanded its mission from the singular objective of flood control to include ecosystem restoration. Budgets for flood-control projects have declined in recent decades, yet the range of objectives for water projects has broadened as society emphasizes environmental and recreational values. Thus, the Corps' portfolio of water-resources projects has changed considerably, with a reduced emphasis on traditional flood-control projects and an increased focus on environmental-restoration projects using nature-based solutions.[23] According to the Corps, the purpose of its ecosystem restoration activities is to restore significant ecosystem function, structure, and dynamic processes that have been degraded, often through its own previous activities.[24]

The term "engineering nature" implies designing nature-based solutions that harness the power and sophistication of nature to turn environmental, social, and economic challenges into innovation opportunities. The solutions aim to help society address a variety of such challenges in sustainable ways—actions inspired by, supported by, or copied from nature.[25] Nature-based solutions use the features and processes of nature, such as the ability to store carbon and regulate water flow to achieve desired outcomes, including reduced disaster risk, enhanced human well-being, and increased access to nature.

After significant effort to control and manage the South Platte River, the Corps, MHFD, and local governments are now investing in nature-based engineering solutions to benefit the local communities and improve river health. Nature-based engineering solutions follow four principal goals for restoring streams in the urban environment:

1. Restore degraded rivers and streams
2. Enhance sustainable urbanization
3. Develop systematic mitigation strategies
4. Improve resilience and reduce risk

These four goals suggest possible high-level actions that the Corps and local agencies, like the MHFD and the City and County of Denver, could implement to incorporate nature-based solutions into river engineering projects. Instituting nature-based solutions increases infrastructure resilience while improving the community well-being in urban areas through natural resource sustainability. To be successful, however, they must be adapted to local conditions and rooted in resilient and adaptable ecological systems.

The Corps began the engineering-with-nature initiative to intentionally align natural and engineering processes to efficiently and sustainably deliver economic, environmental, and social benefits through collaborative processes.[26] The Corps now relies on natural processes and community support to sustain human life and well-being through ecosystem services rather than the "develop, advance, and defend at all costs" strategy it endorsed for more than a hundred years.

The transition from the flood-control era—characterized by discipline-bound, top-down practices—to an interdisciplinary, practitioner-focused process, such as engineering with nature, resulted in a paradigm shift within the Corps. In turn, this has led to a replacement of one-size-fits-all solutions with a geographical focus on place-based contextualized explanation across systems and values. Through innovative application of scientific and engineering principles to design, construct, and implement efficient natural systems, Corps projects benefit from collaborations and partnerships across perspectives, disciplines, and organizations to deliver better outcomes. In practice, the engineering-with-nature concept adds a conservation-focused element to river management, equally weighing ecosystem goods and services (i.e., water-supply and flood-hazard mitigation) with ecosystem function.[27]

The engineering-with-nature initiative in the United States—led by the Corps and flood-control agencies like MHFD—is pursuing a vision for water infrastructure based on this intentional alignment of place-based social context and nature-based engineering processes to efficiently and sustainably deliver economic, environmental, and social benefits through collaboration.[28] The following section provides descriptions of real-world examples where the Corps, MHFD, the Greenway Foundation, and local communities partnered to engineer nature to return natural and beneficial functions to the river.

If We Build It

The highly degraded and channelized South Platte provided an excellent opportunity to implement the Corps' engineering-with-nature initiative. The City and County of Denver, in cooperation with other local stakeholders and federal partners like the Corps, have invested tens of millions of local dollars into the South Platte River Corridor over the past decade. The *River Vision Implementation Plan*, Pelican Ponds Open Space, and River Run Park demonstrate how engineers have changed their approach from structural-based engineering practices to nature-based solutions coupled with the community planning and engagement processes. These projects also illustrate how to engineer nature in order to revitalize a river, including restoring its riparian and aquatic habitats. Ultimately, the implementation of "softer" river-engineering solutions focused on restoration and health rather than traditional flood-control strategies.

River Vision Implementation Plan

A thirty-five-year collaboration between private and public partners resulted in the creation of the South Platte River Greenway, which serves as a main transportation and recreational link for adjacent neighborhoods.[29] The Greenway Foundation, established by Joe Shoemaker and run by his son Jeff since 1982, has worked to improve that link through purposeful redevelopment of the South Platte River. Leveraging collaborative planning efforts between 2006 and 2010 with the Corps, MHFD, the City and County of Denver, and many other stakeholders and community leaders, the Greenway Foundation prepared two master plans for the river corridor within the City and County of Denver: *The River North Greenway Master Plan* (2009) and *The River South Greenway Master Plan* (2010). Jeff Shoemaker compiled the recommendations from those two plans into the *River Vision Implementation Plan* to prioritize project funding. *River Vision* was an ambitious plan and would transform the blighted South Platte into Denver's greatest natural resource. It separated projects into a series of short-, mid-, and long-term priorities to open doors for funding. By developing a single-implementation and priority-setting document, *River Vision* focused on enhancing the health of the South Platte through improvements to aquatic habitat, public open space, aesthetics, water quality along the riverine corridor, and recreational, environmental, and flood control. Novel engineering concepts were nearing reality.

The City and County of Denver, along with the Greenway Foundation and MHFD, collaborated with engineers, landscape architects, and urban planners to develop and implement *River Vision*. Specifically, the plan recommended multipurpose improvements spanning floodplain reconnection, such as instream aquatic habitat, channel, and bank modifications. The plan also included park and recreational improvements like nature playgrounds for unstructured play and surf waves. The river benefited from modern approaches in urban-stream restoration, including:

Creation of a river-focused urban environment
Promotion of ecosystem restoration
Enhanced accessibility, safety and security, aesthetics, and water quality
Creation of a parks/open-space-oriented model for urban living
Maintaining or improving flood-control capacity and stability of the river

River Vision required stakeholder cooperation and compromise to deliver desired results—connecting the people to their river. This required creative and innovative solutions to satisfy the varied objectives of multiple stakeholders. MHFD and Denver Parks and Recreation asked tough questions about the resiliency of each project. Everyone learned together how the river would react to natural and nature-based solutions Through creative solutions, *River Vision* delivered valuable enhancements to the river ecosystem, public health, safety, welfare, and enjoyment of the surrounding environment. Components of the program served as models for similar successful riverine-corridor improvements in metro areas and catalyzed the Denver Urban Waterways Study. It also pumped more than $30 million of restoration projects into a ten-mile stretch of the South Platte in Denver. Rehabilitated riverside parks along the South Platte have reengaged citizens as the river provides recreation while maintaining and, even in some cases, improving flood conveyance. The floodplain provides security for roads, houses, and businesses, while parks provide recreation opportunities for people adjacent to the river. Near-river locations have become popular for stadiums, residences, and businesses. Those investments in the South Platte and its tributaries have facilitated more than $10 billion in residential and commercial development throughout the Denver metro area.[30] This estimate does not include the additional dollar value of air and water quality and public health benefits from green infrastructure. Ultimately, the successful outcomes of *River Vision* can be used in the development of parks and open-space-oriented models for urban living along river corridors across the country, customized to fit the history and character of each location, as along Denver's greenway corridor.

Pelican Ponds Open Space

On a warm summer day in 2019, the Pelican Ponds Open Space in the city of Thornton, Colorado, officially opened to the public. Pelican Ponds Open Space is a two-hundred-acre former sand and gravel mining site within the floodplain of the South Platte River. The gravel mining resulted in a significantly degraded ecosystem prone to occasional flooding that often damaged the recreational trails. The size of the property, frontage along the river, and the variety of natural habitats (open water, uplands, wetlands, riparian forest) presented a tremendous opportunity for engineering with nature and ecosystem restoration, as well as recreational enhancements. Engineering-with-nature highlights included a secondary channel to improve habitat and increase flood conveyance in South Platte, as well as plantings and seeding of native vegetation of twenty acres. Recreational amenities included picnic shelters, fishing piers, and improved shoreline access, overlooks, and trails. A nature playground was also constructed to encourage children to explore nature through unstructured elements such as logs, tree branches, and stones.

River Run

River Run Park, south of downtown Denver, brought new life to a half-mile section of the river. The park provides flood protection, recreation, and connectivity to neighboring communities. On any given summer day, the park is full of people walking and biking the regional path or surfing the new set of river waves. Ben Nielson, the engineer behind the design of the surfing waves, noted, "This multipurpose river enhancement . . . really illustrates the direction of urban river revitalization across the country."[31] River Run is now Colorado's premier river surfing destination on the South Platte. In addition to providing recreational opportunities, the project improved terrestrial and aquatic habitat throughout the river corridor through water-quality enhancements and fish passage through the river structures using natural river aesthetics.

12. River Run Park in Littleton, south of downtown Denver, represents the creativity and collaborative style commonly applied in engineering with nature projects that involve many stakeholders and funding partners. Courtesy of Naranjo Civil Constructors.

The foresight and skills of Laura Kroeger, engineering manager for the Mile High Flood District, were instrumental in the collaborative effort among many stakeholders that resulted in River Run. Her vision and passion for fostering relationships within the community are a model for engineers working in the public sector. Kroeger noted that "we were very pleased to partner with the other agencies so they could also serve a dual purpose for high end recreation. A project of this magnitude doesn't happen without an amazing group of partners."[32]

A River with a Future

In 2011 the U.S. Environmental Protection Agency (EPA) selected the South Platte River as one of seven urban watersheds to pilot their Urban Waters Federal Partnership. According to the EPA, "The Urban Waters Federal Partnership reconnects urban communities, particularly those that are overburdened or economically distressed, with their waterways by improving coordination among federal agencies and collaborating with community-led revitalization efforts to improve our Nation's water systems and promote their economic, environmental and social benefits."[33] The South Platte River Urban Waters Partnership (SPRUWP) is a collaboration of organizations—including the Corps, the City and County of Denver, the MHFD, and the Greenway Foundation—working across governmental and disciplinary boundaries. "The partnership emphasizes stewardship and community connection, linking urban areas with forested watersheds and people with nature."[34] The SPRUWP recognized that local or federal action alone is insufficient to address the complex challenges surrounding water, resource protection, and connecting people with nature. Through a partnership with more than seventy groups, the SPRUWP protects one of Denver's primary sources of drinking water, while supporting ongoing on-the-ground projects, education, and research, as well as facilitating a growing population's connection to its urban waterways.[35]

The City and County of Denver and the Corps initiated two studies to craft the next twenty-year vision for the South Platte River. The first is the Adams and Denver Counties, Colorado: General Investigation Study—also known as the Denver Urban Waterways Restoration (UWR) Study—and the second is the Southern Platte Valley Ecosystem Restoration Study. In assessing riparian habitat and wetland projects along the South Platte, these projects include engineering-with-nature innovations and ecosystem-restoration initiatives to address the long-term degradation along the South Platte River in Denver. A visionary project born out of the UWR could revitalize Denver and its river for generations to come. The Corps and City and County of Denver initiated the UWR study along the South Platte to address the loss of riparian, wetland, and aquatic habitat associated with the creation and operation of three Corps' reservoir projects (Chatfield, Cherry Creek, and Bear Creek) and overall urban development.[36] The UWR study is the foundation for a comprehensive river restoration effort that encompasses approximately 6.5 miles of the South Platte from downtown to the county line. The UWR identified cost-effective plans to restore habitat and ecological function of the South Platte River. In addition, the study focused on ways to reduce flood risks and improve recreation opportunities as a complement to other project goals.

If constructed, the $392 million UWR projects would implement engineering-with-nature initiatives, substantially expanding habitat restoration on a large scale, ensuring a longitudinally connected river corridor. The Corps believes that this project, when integrated with the other local investments in the river corridor, could help stop the historical decline in habitat and waterfowl, as well as decrease the negative impacts on other species of concerns within the region.

A much-scaled-down version, the Southern Platte Valley Ecosystem Restoration study also evaluated the feasibility of improving the quality and quantity of wetland and riparian habitats, creating habitat connectivity, and incorporating in-channel aquatic refugia along the South Platte in Denver but without reconfiguring existing infrastructure and without impacting the existing floodplain.[37] This study proposes minor structural changes to the river, unlike the UWR study, which would completely reconfigure the channel and portions of the adjacent floodplain. The project area is also roughly one-third the length (approximately 2.4 miles). The recommended plan in the study focused on restoration of the aquatic habitat and restoring the wetland and riparian habitat adjacent to the river. The improvements would result in restoration of approximately 1.7 miles of stream length of low-quality aquatic habitat through the construction of in-stream engineering features such as riffles and secondary channels and approximately 11 acres of riparian habitat and 1.5 acres of wetlands.

Engineering Nature: Strategies and Lessons for Urban Rivers

Urban rivers are often the only discernable part of predevelopment landforms and the last "natural" areas in an urban environment. Different land uses, imperviousness, and vegetative cover produce combinations of stressors that affect, often adversely, the health of an urban river. The many socio-ecological complexities of urban environments confound the management of rivers that flow through these environments, particularly when it comes to finding appropriate interventions to improve the physical condition and ecological integrity. As a result, urban rivers are complex socio-ecological systems, often degraded, yet highly valued by communities. The linkages between degraded river health (altered flow and sediment regimes, poor water quality and aquatic habitat, etc.) and social values (experiential, aesthetic, and cultural attributes) are often opaque, making it difficult to develop, justify, and fund appropriate restoration approaches and mitigation actions. Therefore, revitalization of urban-stream corridors

13. A post-restoration rendering of the South Platte River along the River Mile. Rendering by Wenk Associates for Revesco Properties.

requires a multidisciplinary approach to address the social and ecological complexities.

With river systems in metropolitan areas under increased pressure from human influences, management actions that address specific river health issues will involve trade-offs between different aspects of stream functions and community interests and values. Specifically, urban river restoration projects revolve around four components that require negotiated trade-offs: flood control, river health, recreation, and economic development. In Denver, the local governments, the Greenway Foundation, and the MHFD are pursuing studies and projects that balance those components. Through a renewed partnership with the Corps, there is immense potential to restore the physical structure and function of the river. The South Platte is a classic example of a river that was modified for the worse—straightened, channelized, and stripped of its natural characteristics by the Corps following the 1965 flood—and now for the better through nature-based solutions by locally and federally funded projects. Over the past twenty years, the engineering techniques to address the urban-stream syndrome have changed substantially.

A prominent example project that leverages engineering-with-nature principles is the River Mile. This project will address flooding and recreational and habitat restoration goals within downtown reach of the South Platte. One of six reaches analyzed in the UWR study, reach 5 extends from Confluence Park upstream for 1.3 miles to the confluence with Lakewood Gulch near the Denver Broncos stadium (see fig. 13). The design restores and enhances the ecological functions—both in the river and along the banks with native plant communities that will line the river corridor. As a part of this restoration work, the design also addresses spills in the one-hundred-year floodplain and creates additional opportunities for recreation along the river corridor.

The proposed engineering-with-nature design will restore a reach of the river that has been damaged by past industrial activity to a state of ecological functionality so it can support habitats for various species of wildlife (ecological lift) consistent with the goals of the UWR study. Ecological lift will be achieved through the establishment of diverse native plant communities on floodplain benches, through in-stream improvements that will significantly improve habitat conditions for fish and aquatic macroinvertebrates, and through habitat connectivity via the removal of the barriers at Confluence Park. The project will increase native riparian habitat and wetlands on the floodplain benches, at stormwater outfalls, and within the "dynamic zone" of the active channel. The proposed native habitat creations will introduce a more diverse in-stream setting and will complement and support the future phases of the larger South Platte River restoration project.

The Corps' endorsement of engineering with nature highlights the benefit of nature-based solutions that align natural processes and engineering practices to deliver economic, environmental, and social benefits. Systems-thinking employs collaborations and partnerships across perspectives, disciplines, and organizations to deliver better outcomes. These visioning processes are fundamental to identify community values and key services of the South Platte and the link to river health. Multidisciplinary and transdisciplinary teams—engineers working with landscape architects, ecologists, geomorphologists, and social scientists—were also a key ingredient to reshaping the South Platte and its watershed. The multidisciplinary teams demonstrated the feasibility of implementing multifunctional watershed management and ecosystem restoration with a diverse group of stakeholders.

The health of the South Platte River serves as the natural, cultural, and recreational backbone of the Denver metropolitan community. The challenge is ensuring that legacy continues for generations to come. The investment in the South Platte and its tributaries throughout the Denver area is monumental for a river once considered "treacherous." Federal investment provides significant opportunities to maximize the benefit of years of local restoration efforts.[38] State and local entities, as well as the citizens of Denver, continue to make investments in waterways through two 2018 ballot initiatives, which will bring additional funds to the MHFD and the City and County of Denver to implement parks, rivers and waterways, trails, and open-space projects. This investment in restoring the river corridor will only continue to grow in the future. These projects demonstrate how engineers have changed their approach from structural-based engineering practices to nature-based solutions coupled with the community planning and engagement processes.

The South Platte River will always suffer from symptoms of the urban-stream syndrome due to its setting within Denver. However, restoring natural processes through modern-day ecological engineering and river-renovation practices is realigning the trajectory of an impaired ecosystem so that it may continue to provide natural and societal benefits into the future.[39] Reestablishing many predevelopment characteristics of the South Platte River by engineering with nature serves as a compelling story for restoring urban streams. Moreover, the evolution of engineering practices along the river has created a historical tapestry that provides a guide for successfully restoring nature in urban streams across the western United States.

NOTES

1. Urban Drainage and Flood Control District, *South Platte River, Master Drainageway Planning, Chatfield Dam to Baseline Road, Phase A* (Denver: Denver Department of Public Works, August 1994), II-1.
2. City and County of Denver, "Taming the Treacherous Platte," 4.
3. Michael Farrell, "A Tale of Two Dams," Platte Basin Timelapse, August 11, 2015, https://plattebasintimelapse.com/timelapses/.
4. Wright, "The Nation's Responses to Flood Disasters," 7.
5. Wright, "The Nation's Responses to Flood Disasters," 11.
6. Cole, "Inventing Cherry Creek," 23.

7. "Cherry Creek Dam and Lake," United States Army Corps of Engineers, accessed February 21, 2021, https://www.nwo.usace.army.mil/Missions/Dam-and-Lake-Projects/Tri-Lakes-Projects/Cherry-Creek-Dam/.
8. Alan Prendergast, "The 1965 Flood: How Denver's Greatest Disaster Changed the City," *Westword*, April 29, 2015, https://www.westword.com/news/the-1965-flood-how-denvers-greatest-disaster-changed-the-city-6668119; Walsh et al., "The Urban Stream Syndrome," 707.
9. Wellington E. Webb and Joseph B. Blake, *South Platte River Corridor: Long Range Management Framework* (Denver: South Platte River Commission, 2000), 32.
10. Eschner, Hadley, and Crowley, "Hydrologic and Morphologic Changes," A1.
11. Eschner, Hadley, and Crowley, "Hydrologic and Morphologic Changes," A1.
12. Alston, "Denver's South Platte River," 2.
13. Michael A. Stevens, *Stream Stability Investigation, South Platte River, Chatfield Dam to Baseline Road* (Lakewood CO: Mile High Flood District, November 1983), 15.
14. Urban Drainage and Flood Control District, *South Platte River, Master Drainageway Planning, Chatfield Dam to Baseline Road, Phase A*, VIII-1.
15. Stevens, *Stream Stability Investigation*, 5.
16. Nadler and Schumm, "Metamorphosis of South Platte," 96.
17. Stevens, *Stream Stability Investigation*, 6.
18. *Segment 15 Water Quality* (Boston: Camp, Dresser, and McKee Inc., June 1994), 2–6.
19. Alston, "Denver's South Platte River," 2.
20. Alston, "Denver's South Platte River," 3.
21. Wellington E. Webb, *Imperative 2000: A Vision for the South Platte River* (Littleton CO: South Platte Working Group, 1995), 5.
22. T. Bridges, P. Wagner, K. Burks-Copes, J. R. Vietri, *Natural and Nature-Based Features* (Vicksburg MS: U.S. Army Engineer Research and Development Center, 2015), 1.
23. Panel on River Basins and Coastal Systems Planning, "River Basins," 1.
24. U.S. Army Corps of Engineers, *Civil Works Ecosystem Restoration Policy*, ER 1165-2-501, September 30, 1999, https://corpslakes.erdc.dren.mil/employees/policy/ER/ER-1165-2-501.pdf, 2.
25. "Towards an EU Research and Innovation Policy Agenda for Nature-Based Solutions and Re-Naturing Cities," Slideshare, April 25, 2015, https://www.slideshare.net/juditurquijo/nature-based-solutionsaandrenaturing-cities.
26. Bridges et al., *Natural and Nature-Based Features*, 2.
27. Hillman, "Integrating Knowledge," 1991.
28. *Final Integrated Feasibility Report and Environmental Impact Statement: Adams and Denver Counties, Colorado* (Littleton CO: U.S. Army Corps of Engineers, 2019), 2.

29. *The River South Greenway Master Plan* (Denver: The Greenway Foundation, January 2010), 4.
30. U.S. Environmental Protection Agency, *South Platte Watershed*, August 2017, https://www.epa.gov/sites/default/files/2017-08/documents/south_platte_watershed_final.pdf, 2.
31. Ben Nielson (River Design Lead), in discussion with the author, June 2021.
32. Nielson, discussion.
33. "About the Urban Waters Federal Partnership," U.S. Environmental Protection Agency, accessed March 14, 2022, https://www.epa.gov/urbanwaterspartners.
34. U.S. Environmental Protection Agency, *South Platte Watershed*, 2.
35. U.S. Environmental Protection Agency, *South Platte Watershed*, 2.
36. *Final Integrated Feasibility Report and Environmental Impact Statement*, 1.
37. *The Southern Platte Valley Ecosystem Restoration Study* (Littleton CO: U.S. Army Corps of Engineers, September 2018), 9–10.
38. *Final Integrated Feasibility Report and Environmental Impact Statement*, 2.
39. Smith et al., "Urban Stream Renovation," 369.

BIBLIOGRAPHY

Alston, Marc. "Denver's South Platte River Corridor Project." *American Society of Civil Engineers* (June 1998): 1–6.

City and County of Denver. "Taming the Treacherous Platte." *Denver Municipal Facts* 10, nos. 1 and 2 (January–February 1927): 4.

Cole, Erin. "Inventing Cherry Creek." *Colorado Heritage* (November–December 2012): 22–25.

Eschner, Thomas R., Richard F. Hadley, and Kevin D. Crowley. "Hydrologic and Morphologic Changes in Channels of the Platte River Basin in Colorado, Wyoming, and Nebraska: A Historical Perspective." *Geologic Professional* (1981): A1–A39.

Hillman, Mick. "Integrating Knowledge: The Key Challenge for a New Paradigm in River Management: New Paradigm in River Management." *Geography Compass* 3, no. 6 (November 2009): 1988–2010.

Nadler, C. T., and S. A. Schumm. "Metamorphosis of South Platte and Arkansas Rivers, Eastern Colorado." *Physical Geography* 2, no. 2 (1981): 95–115.

Panel on River Basins and Coastal Systems Planning. "River Basins and Coastal Systems Planning within the U.S. Army Corps of Engineers." *National Academies Press* (2004): 1–183.

Smith, Robert F., Robert J. Hawley, Martin W. Neale, Geoff J. Vietz, Erika Diaz-Pascacio, Jan Herrmann, Anthony C. Lovell, Chris Prescott, Blanca Rios-Touma, Benjamin Smith, and Ryan Utz. "Urban Stream Renovation: Incorporating Societal Objectives to Achieve Ecological Improvements." *Freshwater Science* 35, no. 1 (March 2016): 364–79.

Walsh, Christopher J., Allison H. Roy, Jack W. Feminella, Peter D. Cottingham, Peter M. Groffman, and Raymond P. Morgan. "The Urban Stream Syndrome: Current Knowledge and the Search for a Cure." *Journal of the North American Benthological Society* 24, no. 3 (September 2005): 706–23.

Wright, James M. "The Nation's Responses to Flood Disasters: A Historical Account." *Association of State Floodplain Managers* (April 2000): 1–116.

9 Cyborg Salmon and Plumbed Basins

Ambivalence and Compromise on the Deschutes

Kirsten Rudestam

The Deschutes River materializes from basalt springs in southern Oregon, collecting Cascadian glacier melt as it flows north through Central Oregon to join the Pacific-bound Columbia River. As arid Central Oregon's largest waterway, the Deschutes carries not just water but histories of human and nonhuman survival. In the 1800s, after the forced relocation of Indigenous inhabitants, the federal government funded large-scale irrigation projects in the watershed to attract Euro-American settlers to the desert landscape. The plans succeeded; by 1902 irrigation prospects in the Deschutes River Basin were among the best advertised in the nation, with marketing strategies describing the Deschutes (translated as "river of the falls") as "the river of gold." Irrigation claimed 90 percent of the river's streamflow, making the desert bloom but expatriating native fish runs and decimating aquatic ecosystems in the process.[1]

Bend is Central Oregon's largest city, nestled in an S-shaped crook of the Deschutes River pioneers had designated "Farewell Bend"—one of the few fordable points on the river. While timber and livestock industries marked the early days of Bend's colonial settlement, the city now typifies the "New West"; its economy is no longer dominated by resource extraction but by tourism, retail, and high technology.[2] The old lumber mill sports a high-end shopping plaza, with retired smokestacks emerging from the REI outdoor store. Brooks-Scanlon, one of Bend's largest corporate sawmill investors, changed its name to Brooks-Resources Inc. and sells real estate subdivisions carved from the company's old timberlands, and riverfront parks are located strategically next to irrigation canals, which real estate agents advertise as "seasonal streams."

While the exponential growth of Bend's population in the late twentieth century resulted in the decimation of its salmon, images of the fish grace the city's fountains, walls, and tourist brochures. Although casualties of the industrialization of the region, salmon have retained their emblematic value and associative sense of place, and locals have strongly resisted the salmon's decline. But how might efforts to restore the endemic salmon complicate our understanding of what constitutes restoration (and salmon) in the first place? The Deschutes, like countless basins throughout America, contains fractured ecosystems as well as concerted efforts to avoid extirpating native species. These efforts reflect not a turn to preindustrial landscapes but to what may be considered hyper industrialization—hybrid landscapes that defy idealized notions of nature as existing apart from human influence.

Perhaps the best illustration for such hybridity is the Pelton-Round Butte Dam complex, completed in 1964 and nestled in the Deschutes Valley of the high-desert landscape just north of Bend. Bend receives almost all of its hydroelectric power from this three-dam facility. Round Butte, the largest of the three, impounds the waters of the Deschutes. The concrete arch of the Pelton Dam blooms 204 feet from the bedrock, capturing enough water to create Lake Simtustus—a placid body of water that spans more than four hundred football fields and is often surrounded with RV campers. The slightly smaller Round Butte Dam and its reservoir-turned-recreation site, Lake Billy Chinook, are planted a few miles upstream at the confluence of the Metolius, Deschutes, and Crooked Rivers.

To allow for fish migration, Portland General Electric (PGE) implemented fish passages throughout the complex, but these modifications failed, and in 1968 PGE abandoned all pretenses of accommodating fish, severing the historic migration of salmon and steelhead in the Deschutes River system. Thirty years later, in a new world of endangered species litigation, the Pelton-Round Butte Dam complex came up for federal relicensing. Municipal agents, Warm Spring tribal members, irrigators, dam operators, and environmental representatives from Bend spent years in meetings to hatch a plan to restore fish passage in this highly fractured river system.

Given the complexity of this particular confluence, this consortium of stakeholders confronted a formidable task. The Round Butte complex sits on the Deschutes directly below the confluence of three rivers: the Deschutes, the Crooked, and the Metolius. Migrating anadromous fish (salmon and steelhead) follow temperature and water currents to find their way back to their hatching grounds. But the stalling of three rivers created obstacles for the fish in their attempts to make the upstream passage. The cold water from the Metolius River flowing into the Deschutes sank to the bottom of Lake Billy Chinook, while the warmer waters from the Crooked and the upper Deschutes moved over the top of the cold currents and either headed back up the arm of the Metolius or swirled around in eddies. Even without the concrete blockade plugging upstream passage, the unpredictable water flows generated by the dams left the lake with no surface currents for migrating smolts to follow downstream as well as confused all attempts of fish to find their way home.

The solution? The Tribes of Warm Springs and PGE came to an agreement to co-own and operate the dam complex, and they designed a $130 million, 273-foot underwater tower that alters the water currents in the lake, directing fish into an underwater elevator that then shuttles them above surface into a sorting facility. From there, larger fish are returned to the lake to be caught by tourists and tribal members, and smaller fish (propagated in an upstream hatchery) are sedated, marked, then released into the lower river, where they continue their migration to the ocean.[3]

The fish migration project was completed in 2010. Politicians from around the country attended the ribbon-cutting ceremony and former Secretary of the Interior Gale Norton praised the project at the tower's inauguration, remarking, "With sound science, cutting-edge technology and creative solutions, we can have both healthy rivers and thriving communities."[4] Thousands of salmon and steelhead have now passed through the dam complex on their way to the ocean, and in 2012 the first adult salmon from the first smolts returned upstream, generating hopes for what engineers believe will be a self-supporting population of anadromous salmon and steelhead.

Previously, in 2001, engineers, tribal representatives, environmentalists, and others came together with a shared goal—to restore the Deschutes

14. Computer-generated model of the underwater tower and fish-collection station at Round Butte Dam. The structure is planted in the floor of Lake Billy Chinook and is as tall as a seventeen-story building. Courtesy of Portland General Electric Company.

River's historic salmon run. But the elaborate revitalization of the salmon does not just demonstrate the prowess of modern technology—its presence and celebration raise questions about how we understand and define restoration in times of significant anthropogenic change and disruption.

The Birth of the Bend Metropolis

In the American West, the combination of settler colonialism and capitalist projects such as railroads, lumber processing, and the doctrine of prior appropriation generated new ways of interacting with local waters and landscapes.[5] Prior to the arrival of Euro-Americans, three different tribes had developed societies in the Deschutes watershed—the Wasco, who had taken up residence in the east, the Walla Walla (later called the Warm Springs), who moved between summer and winter villages along the Columbia River, and the Paiute, who lived in the southeastern portion of the basin. Despite eventually being forced into one shared reservation south of the Columbia on the Deschutes, Metolius, and Warm Springs

Rivers, these bands of Indigenous peoples are and were distinct, with different lifestyles, practices, and languages. However, each maintained a respectful and sacred relationship with salmon and the waters through which they swam.[6]

In the late 1800s, after the federal government forced the remaining Indigenous residents onto a shared reservation, the railroads launched a campaign to fill the now-emptied landscape, pitching the Deschutes as a place of natural wealth and abundance. But settlers were often dismayed by the unanticipated hardships obscured by the advertisement of plenty. Unlike other parts of the Pacific Northwest, the arid climate and the short growing season of Central Oregon deterred immigrants, whose efforts to farm the thin volcanic soil yielded little return.[7] The U.S. government attempted to attract settlers to the area, reporting that the region contained "grazing country sufficient for numbers of flocks and herds," but five years later, Capt. John Drake issued a contrary impression: "As for the country, I have no desire to visit it or any portion of it again. It is a desert to all intents and purposes, utterly worthless, sandy, rough and rugged . . . with a stunted growth of juniper covering the surface." The few hundred settlers who did arrive in the area in the mid-1850s organized themselves into loose communities adjacent to reliable water sources, and most subsisted by raising livestock.[8]

Two infrastructure developments catalyzed dramatic changes in the Deschutes landscape toward the end of the nineteenth century: the railroad and irrigation projects. During the first several decades of Euro-American settlement, much of Oregon and Washington had become recognized for high-quality timber. Areas with access to river shipping on the Columbia quickly grew, but because of its relative distance from markets, Bend and the surrounding area relied primarily on subsistence practices and small-scale trade until the completion of the Deschutes railroad in 1911.[9] The Oregon Trunk Railroad line connected the region to national and international markets and expedited large-scale timber production from Central Oregon forests, sparking a fierce competition between two lumber companies—Brooks-Scanlon and Shevlin-Hixon—who built mills on opposite sides of the river and raced to harvest the largest trees.[10]

While timber processing enjoyed high productivity, farmers were having a difficult time living up to the Jeffersonian ideal of self-sufficiency in the dryland of Central Oregon, attempting to eke out fruits and vegetables from climate and soils not well suited for agriculture. Irrigation was the solution. Settlers with access to surface water dug ditches and canals to irrigate their small plots of land, but these small-scale operations were shadowed by the advent of large-scale federally funded irrigation projects. By 1920 $12.7 million had been invested by shareholders who subsidized private or shared irrigation projects, and federal irrigation policies sped through efforts to bring water to more than 986,000 acres of previous desert lands.[11]

By the 1950s the frenetic pace of tree removal and wood processing had slowed down. Most of the lucrative timber had been harvested, and irrigation districts had gobbled up the water rights to all of Central Oregon's surface waters. In the 1970s a new form of land-based profit emerged in conjunction with economic restructuring. As primary industries were outsourced overseas, capitalist initiatives concentrated on other services, such as recreation, retail, and professional services.[12] Far from its origins as a small town reliant on natural resource extraction, the city of Bend now embodies the New West—a region characterized by "a burgeoning winter sports tourism industry, the emergence of telecommuter communities, and the migratory phenomenon from urban to rural environments crudely referred to as 'white flight.'"[13] An increasingly vocal and powerful constituency now urges lawmakers to reallocate water not for farming but for recreation, tourism, and ecological integrity—all of which brings us back to the elaborate efforts to regenerate viable habitat for the endemic salmon.

Fish without Rivers

In 1960 the Washington Department of Fisheries boasted that new methods of "salmon egg incubation [and] predator and hydraulic control in water areas, plus the impoundment of migrating salmon at or near the rearing ponds for the artificial taking of spawn" could now provide "salmon without rivers."[14] Thirty years later, CTWSRO (Confederated Tribes of Warm Springs Reservation of Oregon) and PGE's expansive and expensive fish tower project would cast a shadow over the pride that Washington's engineers had for their 1950s' hydraulic initiatives. The fish tower project demonstrated a feat of technological imperialism; while the salmon still enact the inexplicable journey of running back up the very same rivers from which they hatched, the rivers bear little resemblance to those that supported the spawning season before the era of dams, turbines, and fish towers.

The elaborate revitalization of a salmon run, such as at the Round Butte Dam, requires two elements: enough water for passage and no obstructions along the way. PGE and the tribes maintain responsibility for the second requirement—development and maintenance of the infrastructure to enable salmon passage. Despite the expense and engineering devoted to fulfilling that task, the first element—ensuring enough water for salmon to swim—may be even more formidable, thanks to the calcified politics surrounding private property and water rights. Wrestling water away from water-rights holders to provide water for fish and rivers required the formation of a new entity—the Deschutes River Conservancy (DRC), a nonprofit organization formed in 1992 by CTWSRO, the Environmental Defense Fund, and local irrigation districts. The DRC is tasked with the mission to "restore streamflow and improve water quality in the Deschutes Basin," and they, like PGE, have resorted to unique methods to reallocate water in the basin.[15]

In the process of creating a river that can host fish, the DRC has had to participate in "waterworld." Employees know that the basin, for all its natural-seeming beauty, is really a plumbed system from the headgates of the dam on Wickiup Reservoir down to the Pelton-Round Butte Dam complex. A network of irrigation canals and diversions lies in between, and the water pumping through those veins and arteries is constantly monitored and managed by the Department of Water Resources. In order to wrestle water from senior rights holders back into the mainstem of the Deschutes, the DRC primarily operates via market-based mechanisms, providing irrigators with more efficient water-transport systems and then paying them for their saved water, which they then leave instream for the fish (a win-win for the irrigators, although many are still resistant to this offer, thanks to a long-standing reluctance of irrigators to forfeit water rights for environmental interests).

The DRC's water transfers are not simple. They resemble a giant natural resources chess game, with units of water and water rights moving around the board in elaborate exchanges. For example, one scenario involves Central Oregon Irrigation District (COID) transferring 2,500 acres of urbanizing lands onto 2,500 acres of North Unit Irrigation District (NUID) Crooked

River lands. In exchange for receiving COID's water rights, NUID transfers its Crooked River rights instream. COID is reimbursed for this arrangement by being paid to conserve 5,000 acre-feet of water by piping unlined canals, also giving them an opportunity to create a hydropower facility for future revenue. Swalley Irrigation District then transfers 250 acres of water rights to the river for mitigation in exchange for payment from DRC. And viola! Three hundred CFS are back in the Upper Deschutes.[16]

Counting and trading water has paid off, and thanks to the DRC, water has made its way back into the river and tributaries during the winter months for the first time in decades, making it possible for the hatchery salmon to survive the trips up and down the river through the new dam complex. Residents and state officials laud these new innovations for providing "healthy rivers and thriving communities." But what are the consequences of creating "fish without rivers"? What do we lose when we equate the technological sophistication of a fabricated fish run with the wild rivers of the past?

Expert Knowledge and Disenchantment

Those actively involved in water politics referred to the basin as being "plumbed." One spring day I had the opportunity to witness this plumbing of the river firsthand when I accompanied the watermaster from the Water Resources Department to Wickiup Reservoir, the main storage facility for Deschutes irrigation districts' water. A watermaster (the official title) is appointed to allocate the legal right to use water from irrigation projects and diversions, as well as penalize those who illegally divert water to which they are not entitled. Each spring when irrigation season begins, the watermaster opens the headgates to the reservoir to release water for irrigation demands. When he asked if I wanted to take a turn at the wheel, I jumped at the opportunity. It was more difficult than I had imagined. Each turn of the heavy wheel made me gasp with effort, and because the water release happens slowly, I did not get to see a dramatic gush of water emerge from between the steel plates. By that time, I had spent many months exploring the Deschutes. I swam in it, wandered along its banks, picked cattails, and spotted great blue herons. I'd bicycled along its paths and watched the fish catch flies at sunset. I'd seen kayakers take advantage of the surfing wave just north of downtown. I had tracked the rise and fall of the river and followed the irrigation canals that crisscrossed through town diverging from the mainstem of the Deschutes. And now I was playing an active role in its flow regime.

The headgates of Wickiup are upstream of the major urban areas, and, as such, the intensive management of the watershed is mostly out of sight of the public eye. In this regard, the plumbing of the river operates as a form of elite or expert knowledge privy to those working in waterworld. For example, an Oregon Fish and Wildlife Service employee described the hidden factors in the flow regime.[17] He stated, "Most people just don't see it. I think most people looking at the river don't see a modified system . . . they don't get that everything they see in front of them is all controlled by rebar, gates, and valves."[18]

Over the course of my time in the Deschutes talking with locals, I found his description of "most people" accurate. I rarely spoke with anyone who described the Deschutes as intensively managed. One long-time Bend resident took me to one of her favorite spots to hike along the Upper Deschutes River. The woman observed the high flows and observed, "Wow, all this hot weather must be melting the glaciers. The river's so high!"[19] I may have suspected the same had I not been with the watermaster a week earlier wrenching open the steel headgates. Even having had that experience, I still found it difficult to wrap my head around the river's bizarre flow regime, where the flows increase in the summer and slow to a trickle in the winter. It went against everything I thought I knew about rivers. A few weeks later, I overheard a DRC staff member describing a recent rafting trip made possible thanks to the high flows, and without thinking I asked, "Because of all the snow melt?" "No," he replied, "irrigation demands." Embarrassed, I said, "Oh right, I should know better," to which he responded, "Yeah, nothing is natural on this river."[20]

Water managers expressed pride in their water marketing and management strategies—strategies that managed to defy the logic of western water policy and return water to the mainstem of the river. But I often heard water managers also express a sense of disenchantment in their work. For example, one day I accompanied the water manager on his rounds to hunt down illegal water diversions. At one point in our journey, he stopped and pointed to a mass of lumber that someone had placed in a tributary to plug and divert the water toward their hay field. "Look at that! That's practically five cubic feet per second [cfs] that they're [illegally] diverting! That's a lot of water."

When I asked him how he became so skilled at seeing water in cubic feet per second, he shrugged, "That's just how I think about it now."

"What about when you're knee deep in the water in a beautiful place?" I asked, knowing that he spends quite a bit of time wading into the river in waders to measure the river flow.

"I wish I could say, 'Yes, I feel differently,' but actually I don't," he responded. "I still see it in terms of cfs."[21]

That this man wished he could say "yes" speaks to a certain inner conflict that many other water managers alluded to as well. One DRC staff member described her shift in perspective that occurred after working with the DRC: "I now see the river as heavily modified," indicating that this diminished her enjoyment of it. A field worker with the Upper Deschutes Watershed Council (UDWC) echoed this perception: "I experience [the river] as plumbed . . . every fish that enters this basin goes through a selection facility and is tagged and then every fish that comes back gets trucked around. . . . Every drop of water in this basin is managed, and I guess I don't see that changing. . . . It's fun to be on rivers that aren't dammed. This just isn't one of them."[22]

The director of the UDWC referred also to a sense of loss or disenchantment in coming to terms with the highly modified plumbing of the basin: "Personally, there are few segments [of the river] that I go to that don't make me sad because of [the human impact]. Every day I think about it. . . . It's hard for me to see a segment of river without seeing how it's been hammered. . . . I think ignorance is bliss for a lot of those folks [who recreate on the river]."[23]

Communicating Restoration

These responses from waterworld managers touch on important themes regarding human relationships with the more-than-human world. For one, they allude to what may be a prevalent attitude among western environmental advocates who, in order to engage in contemporary restoration work, must confront their idealized visions of a human-free nature. As previously noted, although they cynically referred to the Deschutes watershed as plumbed, many of the water managers I talked with acknowledged their own tendencies to value that which is "natural" over that which is technological, thus reiterating this familiar nature/society polemic. That said, the majority of rivers in the American West have been dammed and diverted for irrigation, power generation, and flood control. The fracturing of these riverine systems has only been exacerbated with the advent of severe droughts and increasing urban pressures. What does "restoration" mean in the context of such disrepair? Not only must we reckon with the undeniable impact of human exploitation, but restoration and repair in this case complicate traditional narratives of nature as that which is wild and untarnished by modern human technologies.

The trouble with reiterating a dichotomy of nature and society/technology is that it doesn't give us a good framework for defining restoration in the context of contemporary anthropogenic change (nor does it give us an accurate or just rendition of history, by rendering invisible the stewarding and habitation of land by Indigenous peoples' precolonial settlement). Water managers spoke to this conflict, and torn between the act of defending and dismantling public perceptions of attachments to "wild" nature, they expressed ambivalence around how best to gain public support for their initiatives. For one, DRC staff want people to understand how compromised the river has been thanks to human modifications, underscoring the plumbed aspects of the waterscape. On the other hand, they also want "people to understand waterworld [the technocratic] without losing the enchantment factor."[24] To this end, the DRC's brochures display glossy photographs of bubbling springs and lush rivers, not dam complexes and irrigation canals, even though the latter are essential to the agency's water-marketing strategies. As DRC's communications strategist stated, "Lay people can wrap their heads around simple messaging. . . . The work we do here is very technical, but it's how we connect to general people [that matters]."[25] And part of connecting to general people is, in this case, assuming that they value the "natural" over the "technological."

Restoration and Compromise

In acknowledging the trouble with applying the nature/society dualism to contemporary water restoration, several water managers resisted the temptation to perpetuate such a dichotomy in their work with the public. Instead, they acknowledged the impracticality of upholding the vision of a human-free nature as a guiding principle in a context where cutting, clearing, relocating, introducing, and otherwise modifying the earth have created irrevocable changes in the waterscape. A DRC project manager asked, "What are you restoring to? Yeah, you live in the land of compromise. We have what we have to work with. The plumbing question—we can't go back to the natural hydrograph. So what can we get from the altered hydrograph?"

Likewise, the director of the UDWC described the goal of his work as not to restore the river to its "original" floodplain and flow regime, but to "create ecological conditions in the river so that it can repair itself in the long run . . . get it to the place where there is sufficient floodplain, where it can support insects, vegetation, get it to a place where it can take care of itself."[26] These responses help demystify and deconstruct our dualistic visions of nature, and describe a river that is both active and passive, reliant on human intervention but able to "take care of itself."

Back at the salmon elevator, the bred, shuttled, and sifted salmon also point to new ways of understanding and defining restoration. Despite the seemingly dry and disconnected practice of counting fish, many biologists, citizens, and dam operators expressed awe, delight, and pride in contacting—via sight, story, or direct touch—the salmon who managed to journey upstream. The salmon salvaged by such feats of technology were not simply salmon as commodities; the sockeye clipped, tagged, and shuttled through the dam complex maintained a charismatic appeal, even if that appeal was hinged upon a nostalgic memory of a once ecologically intact watershed. Photographs of that sockeye salmon circulated through news articles and press releases. One PGE representative exclaimed, "It's pretty exciting. . . . It's like our first grandkid just got to college. This is what we all dreamed about."[27]

Likewise, Warm Springs tribal members I interviewed celebrated the PGE dam passage project while revering the sacred cultural dimensions of both water and salmon. During one visit to the dam complex, I encountered the natural resources director for the Warm Springs Tribes. He shared with me the news from that year's autumn run. "The first sockeye made its return!" he exclaimed. "Ta-da! It was with great fanfare. The first adult fish." He said that everyone on the reservation knew about it right away. I asked him if tribal members felt conflicted around the dam's operations, and he responded matter-of-factly, "Sure, the naturalness of the system has been corrupted. Back in 1958, with Bonneville Dam they created lake habitats, so species have adapted to this system, changed temperatures, changed the flow regime. You have manmade problems, so you have to have manmade solutions."[28] Clearly, these "manmade solutions" did nothing to dampen his enthusiasm for the return of the salmon run. Instead, he expressed an intimate fondness that was echoed by others I interviewed.

For example, during one tour of the facility my engineer guide frequently empathized with the "poor little guys" that get wrung around in the gears of the contraption, and he said that one of his main contributions to the facility was a fish-handling procedure that reduces fish harm and attrition. He reported with pride that thanks to his innovations, the fish were treated as gently as possible. As we walked, he pointed to the fish circling around in the holding tanks, and he paused for a while, peering in with curiosity and thoughtfulness, indicating both a pride in and an appreciation for their size and vitality. At one point he warned me to avoid the electric wiring that bordered the various fish pipes and tanks. Apparently before implementing the electric fence, local critters discovered that easy meals could be secured by visiting the fish facility. The engineer said that great blue herons, mountain lions, and raccoons were all frequent visitors, but he was particularly offended by one creature—the river otter—who broke into the facility and used the fish-holding tank to teach its young how to kill fish. "Within minutes," he told me with visible sadness and frustration, "it had killed about one hundred fish!" that it left uneaten and scattered about the facility. In recounting this incident, he demonstrated a care for the fish generated over the years of intimate contact.

Restoring to What? Restoring to Whom?

While conflict around water is perhaps de rigueur in the American West, the expressions of ambivalence, hubris, and accommodation with respect to restoration indicate that water managers are actively grappling with how to make sense of and relate to restoration in a world of compromise. Those I interviewed searched for ways to reconcile their desire for environmental resilience with the recognition that there is no going back to a non-modified "nature," describing the erosion of their natural images in the process of achieving their management goals. Public-relations efforts in the Deschutes revealed their inner conflict; water managers found themselves strategically positioning different versions of the river (natural vs. technologically controlled) to gain public acceptance and recognition for various projects.

The Deschutes illuminates the ways in which the frames of nature and of technology exist as cultural ideals, mediating river-restoration practices. At the same time, managers described river restoration as a terrain of compromise, where effective work entails realistically addressing the accumulated human impact on riparian systems over time. Historian Richard White describes in *The Organic Machine* the evolution of the Columbia River from a life-affirming and life-giving entity to an energetic process fractured by claims for irrigation, electricity, conservation, and commercial fish industries. In charting the course of the river and its kinetic momentum downstream, he resists reinscribing the binary of nature and society onto the Columbia. The question that he poses to his readers is how we might see the river as a coherent whole rather than a set of separate parts utilized for different stakes. "Some," White writes, "would reduce the consequences to a cautionary tale of the need to leave nature alone, but to do so is to lose the central insight of the Columbia; there is no clear line between us and nature."[29]

Thirty years later, stories from the Deschutes resonate with White's argument. If we consider its salmon, we see how their very existence is contingent upon a web of wheels, dams, and pulleys, making it impossible to retain the idea of nature as a realm devoid of human social relations. This salmon can be considered a cyborg, a living organism that relies on human technology for its survival. What can such salmon (and "organic machines," such as White's Columbia River) teach us about our relations with the more-than-human world? Cyborgs confuse the boundaries between science and nature; they demonstrate how humanity has become so imbricated into technology (and vice versa) that we can no longer divide the line between human and machine.[30] In breaching these boundaries between society and nature, we are forced to acknowledge and admit to complexity. If salmon, rivers, and even our own selves are tangled networks of society and nature, how might this perspective aid us in responding to White's central insight? Based on the conversations I had with water managers, it may be that acknowledging such hybridity, and with it such compromise, might help us develop more adaptive and creative orientations to concepts and applications of restoration and resilience.

In suggesting that we honor and connect to the hybrid landscapes that are our home, I do not intend to ignore the unjust and critical impact of capitalist exploitation on western waterscapes, nor make light of all that we have lost in the mad frenzy to claim resources from the land. Climate disruption, forest fires, acute water stress, cultural fragmentation and dislocation, and extinct and endangered species are only a handful of the dire consequences resulting from humanity's unethical relationship with the earth and each other. And I do believe that we lose something when what used to coexist within a rich ecosystem of wild rivers, human ceremony, black bears, and old-growth trees becomes extracted from these relations.

It may help to remember that despite the mirage of a nature devoid of human influence, the nonhuman is not inert matter waiting to be molded by human technologies. Nature has its own agency; the movements and motivations of rivers, rocks, storms, and salmon are often unpredictable, capricious, and even baffling to western science. The Deschutes River has been aptly referred to as "peculiar" by hydrogeologists.[31] Thanks to the river's unique morphology, early initiatives to create reservoirs in the porous basalt resulted in failure, and even the celebrated fish tower broke in half midway through construction thanks to unanticipated river currents. Even biologists have never quite understood the Deschutes salmon, who manage to find their pathway back to the rivers of their birth.[32] White speaks of the Columbia in similar terms; while ruptured and modified, its organic momentum represents a force beyond human ingenuity and technological prowess. When we consider the hybrid nature of such waters, it may serve us to remember (and temper our hubris) to recognize all that remains wild and undomesticated within our local waterscapes.

And what about the iconic salmon? My tours with engineers and conversations with tribal members suggested that technology might not always sully our experience and relation with nature but at times enhance it. The success of the fish-passage project is measured by the "yield" of returning fish, and the one central purpose of the facility is to capture, sort, and count. But while the ecology and life worlds of local fish can be reduced to body counts, my time spent with those experientially engaged with the fish reintroduction project illuminated how contact with fish engendered sentiments that alluded to surprisingly vehement relational intimacy in the form of protectiveness and care.

Sociologist Hillary Angelo explores how our understanding of nature is formed through our intimate experiences, claiming that "experience is a door to empathy."[33] In the Deschutes, we have two renditions of salmon: those that are encapsulated in murals, tourist brochures, and Bend's downtown architecture, and those that have been captured, tagged, clipped, and shuttled. While representing very different ideologies, the former, in maintaining a sense of place that is wedded to the iconic salmon, helped mobilize public support for such a grand and costly endeavor, whereas the latter gives contemporary relevance to the tourist brochures and downtown art. And those who have the opportunity to get close to the fish themselves—they may develop an understanding of and appreciation for the critters that do not fit either discrete category. Their salmon are not those that are simply counted and commodified, nor are they just a regional icon. Instead, we see what can happen in the rich expanse of an encounter, when we get close to our nonhuman others and build relationships that honor hybrid natures. If we save our love for that which does not exist, we may lose our opportunities to develop affective ties with and learn to care for our earthly, more-than-human homes that are comprised of technology and nature, the wild and domesticated. And in reevaluating what we understand to be nature, perhaps we can regain and retain an intimacy with and affection for local waters and species that are not predicated on them being untouched by human engagement.

And so we are left with the Deschutes—a contemporary river that both dissolves and affirms narratives of nature and wilderness. In its northern journey, it reveals the inextricable social relations upon which its waters flow as well as the unpredictable and unmappable forces of water, fish, and webs of relations that contribute to its peculiar nature. By demystifying and deconstructing dominant dualistic visions of nature, we may perhaps find possibilities for new modes of river restoration in the west, and new pathways of reconciliation between human and riparian worlds.

NOTES

1. Hall, "Irrigation Development," 39.
2. Robbins, "In Pursuit of Historical Explanation," 279.
3. "Restore Fish Passage," Columbia River Inter-Tribal Fish Commission, accessed June 15, 2015, https://critfc.org/tribal-treaty-fishing-rights/policy-support/columbia-river-treaty/restore-fish-passage/.
4. "Collaborations in Indian Country," Hydropower Reform Coalition, accessed May 23, 2012, https://hydroreform.org/resource/collaborations-in-indian-country/.
5. The doctrine of prior appropriation determined that the earliest users of water have the right to take all they can use before a subsequent user can take water from the same tributary.
6. "History of the Tribes of Warm Springs," Wisdom of the Elders, accessed August 2, 2016, https://wisdomoftheelders.org/discovering-our-story-airs-on-november-6/.
7. While the volcanic ridges of the Cascades are bestowed with more than two hundred inches of rain and snow a year, the Deschutes Basin receives a fraction of this amount, averaging ten inches per year.
8. "The Oregon History Project," Oregon Historical Society, accessed December 21, 2009, https://www.oregonhistoryproject.org/.
9. Robbins, "In Pursuit of Historical Explanation," 289.
10. "The Oregon History Project," Oregon Historical Society.
11. Hall, "Irrigation Development," 42.
12. Claeyssens, "Waldo Lake History," 27.
13. Robbins, "In Pursuit of Historical Explanation," 290.
14. Lichatowich, *Salmon without Rivers*, 43.
15. "Mission and Vision," Deschutes River Conservancy, accessed December 13, 2021, https://www.deschutesriver.org/about-us/mission-and-vision/.
16. Rudestam, "Whiskey Is For Drinking," 4.
17. To ensure anonymity, all interviewee names have been deleted.
18. Rudestam, "Whiskey Is for Drinking," 108.
19. Rudestam, "Whiskey Is for Drinking," 108.
20. Rudestam, "Whiskey Is for Drinking," 109.
21. Rudestam, "Whiskey Is for Drinking," 111.
22. Rudestam, "Whiskey Is for Drinking," 112.
23. Rudestam, "Whiskey Is for Drinking," 111.
24. Rudestam, "Whiskey Is for Drinking," 115.
25. Rudestam, "Whiskey Is for Drinking," 122.
26. Rudestam, "Whiskey Is for Drinking," 116.
27. Cassandra Profita, "At Pelton Round Butte Dams: $100 Million Steelhead Returns," Oregon Public Broadcasting, 2013.
28. Rudestam, "Whiskey Is for Drinking," 188.

29. See White, *The Organic Machine.*
30. Haraway, *When Species Meet*, 39.
31. O'Connor and Grant, *A Peculiar River*, 12.
32. Lichatowich, *Salmon without Rivers*, 23.
33. Angelo, "Bird in Hand," 36.

BIBLIOGRAPHY

Angelo, Hillary. "Bird in Hand: How Experience Makes Nature." *Theory and Society Journal* 42 (June 2013): 351–68.

Choy, Timothy. *Ecologies of Comparison: An Ethnography of Endangerment in Hong Kong*. Durham NC: Duke University Press, 2011.

Claeyssens, Paul. "Waldo Lake History: Prehistoric Period to Present." *Lake and Reservoir Management* 16 (June 2000): 25–39.

Cronon, William. "The Trouble with Wilderness: Or, Getting Back to the Wrong Nature." *Environmental History* 1, no. 1 (January 1996): 7–28.

Hall, Michael. "Irrigation Development in Oregon's Upper Deschutes River Basin 1871–1957: A Historic Context Statement." *Bend Historical Society Journal* (August 1994): 1–139.

Haraway, Donna. *When Species Meet*. Minneapolis: University of Minnesota Press, 2008.

Lichatowich, Jim. *Salmon without Rivers: A History of the Pacific Salmon Crisis*. Washington DC: Island Press, 1999.

Moore, Donald, Jake Kosek, and Anand Pandian. *Race, Nature and the Politics of Difference*. Durham NC: Duke University Press, 2003.

Nightingale, A. J. "Beyond Design Principles: Subjectivity, Emotion, and the (Ir) Rational Commons." *Society & Natural Resources* 24, no. 2 (January 2011): 119–32.

O'Connor, Jim, and Gordon Grant. *A Peculiar River: Geology, Geomorphology, and Hydrology of the Deschutes River, Oregon*. Corvallis: Oregon State University, 2003.

Robbins, William. "In Pursuit of Historical Explanation: Capitalism as a Conceptual Tool for Knowing the American West." *Western Historical Quarterly* 30, no. 3 (Autumn 2013): 277–93.

Rudestam, Kirsten. "Whiskey Is for Drinking: Water Passions and Water Politics in the American West." PhD diss., University of California, Santa Cruz, 2017.

White, Richard. *The Organic Machine: The Remaking of the Columbia River*. New York: Hill and Wang, 1995.

10 Neon River

Rethinking Chicago Environmental History and Personhood in the Age of Humans

Shawn Bailey

Nelson Algren often referred to gritty, mid-twentieth-century Chicago as a "neon wilderness." In his 1951 prose-poem *Chicago: City on the Make*, Algren described "a city that was to live by night after the wilderness had passed. A city that was to roll boulevards down out of pig-wallows and roll its dark river uphill. That was to forge, out of steel and blood-red neon, its own peculiar wilderness."[1] Algren's neon wilderness was home to the enduringly down-and-out—forlorn gin soaks and barroom brawlers, prostitutes and pimps, luckless grifters and corrupt cops. Their stories were often dark, sometimes tragically funny, but most offered a small glimmer of hope and the possibility of salvation. The river that bifurcates this neon wilderness is similar in many regards. The Chicago River—a neon river—is a downtrodden ecosystem reconfigured and defiled by human hands for centuries. Yet, like a Nelson Algren character, there remains the prospect of conditional redemption.

Brute force and violence punctuate the history of this neon river. At the end of the Pleistocene, retreating glaciers raked channels in the region's silty soil, and melting ice filled these shallow grooves with water that flowed, almost reluctantly, into nearby Lake Michigan. For millennia, the river proved a strategic and welcoming place for successive human cultures—a watery crossroads with the capacity to connect the continent. During the nineteenth century, the river was a virulent trough of human excrement, decaying meat, cholera bacteria, and waning populations of wildlife. Chicagoans, unwilling to end their mistreatment of the river, proposed an audacious solution to their self-inflicted problem—they decided to roll their river uphill.

Completed January 2, 1900, the reversal of the Chicago River was one of the engineering marvels of its day—a "hydrologic coup d'état" and the first time human beings completely reordered the natural flow of a river. Afterward, Chicago's filth swept toward the Gulf of Mexico via the Mississippi River watershed, and uncontaminated water from Lake Michigan flushed the city's Main Branch, unintentionally beginning the sluggish rewilding of the Chicago River. Today, about sixty species of fish swim in the Chicago River watershed, joined by an impressive menagerie of other animals. Not all thriving species are indigenous, however, contributing to the Chicago River's reputation as an "invasive species superhighway."[2]

How can the environmental history of this neon river help make sense of the active and chaotic nature of Chicago, especially in the Anthropocene? Chicago's rewilding—both intentional and unintentional—should not be interpreted as an effort to restore the river to its "natural" self before the impact of human hubris and the rise of a major metropolis. Philosopher Eric Katz calls this kind of environmental restoration a "big lie," and any twenty-first-century pining for a completely resurrected Chicago River would be buying into one heck of a whopper. Instead, Chicagoland's 9.5 million humans must understand that they are a part of something unique—a new era and a new river. Human beings are an integral component of the river's biology and must make choices that create room for all Chicago species.[3] In doing so, Chicago can take lessons from both Nelson Algren stories and the complex environmental history of the region and envision a bold future that guarantees the legal rights, ecological health, and even the personhood of its neon river.

Greenbelt Forest Preserve is a good place to ruminate on the geology of the Great Lakes region and the human history of Chicagoland. Tucked between the blue-collar towns of Park City and North Chicago, immediately south of the city of Waukegan and some forty miles north of downtown Chicago, Greenbelt is a grassy oasis amid an industrialized landscape. Part of the Forest Preserves of Lake County, the park contains five miles of well-maintained hiking trails—encased by prairie flowers and eight-foot cattails—which weave around small ponds filled with sunfish, herons, and basking turtles. Greenbelt is also the northernmost headwater of the Chicago River.

Like all Chicagoland, Greenbelt is a product of the Pleistocene. Roughly seventy-five thousand years ago, an offshoot of the massive Laurentide Ice Sheet known as the Wisconsin Glaciation covered present-day Chicago in 1,200 feet of ice. This glacier was a dynamic force that carved the Great Lakes and the Chicago River. As this ice slid southward—like a giant sheet of icy, rough-grit sandpaper—it plowed through the silty soil of the region. At its edges, the glacier pushed till into ridges called moraines and created a subcontinental divide between two distinct watersheds. Approximately sixteen thousand years ago, the Wisconsin Glaciation began a slow melt that filled the region's lakes and rivers.[4]

Greenbelt links to the Chicago River because of this ice-age geology. The western boundary, now a bustling thoroughfare of speeding cars and tractor trailers known as Green Bay Road, sits atop a thirty-foot-tall Pleistocene ridge called Blodgett Moraine. Precipitation on the eastern slope of this moraine collects in the marshlands of Greenbelt before traveling south as the Skokie River. The Skokie joins two other streams to form the North Branch of Chicago River, which connects with the South Branch at Wolf Point in present-day downtown Chicago. Until 1900 this Main Branch flowed east into the 279-foot depths of Lake Michigan, through the Great Lakes system, and eventually into the Atlantic Ocean. Water west of the subcontinental divide travels southwest in the Des Plaines, eventually reaching the Mississippi River via the Illinois River.[5] At the close of the Pleistocene, then, Chicagoland ridges, sometimes only a couple of feet tall, separated an intercontinental water system connecting the northern Atlantic Ocean with the Gulf of Mexico. For successive human cultures, this near linkage proved a beguiling opportunity for trade and growth.

The first Chicagoans arrived in the region via Pleistocene pathways and used the Chicago River to their full advantage. Early hunters likely stalked woolly mammoths and mastodons from the moraines of the region, eschewing the swampy bottomland headwaters of the Chicago River. In Kenosha County, Wisconsin, near the same ridge system that links Chicago to Green Bay and less than twenty miles from the headwaters of the Chicago River, researchers excavated two of the most complete mammoth kill sites in North America. In 1993 and 1994, they unearthed largely intact skeletons and the stone tools used to hunt and butcher the carcasses. According to archeologist David Overstreet, evidence collected at these sites "point in the same direction—that somebody in southeastern Wisconsin was mucking about with mammoths 15,000 years ago."[6]

Native Americans have remained in the Chicago area ever since. Numerous Indigenous groups occupied Chicagoland, either in permanent settlements or through seasonal use, including the Delaware, Kickapoo (Kiikaapoi), Illinois, Mascouten, Menominee, Mesquakie, Miami (Myaamia), Odawa, Ojibwe, Potawatomi, Sauk, Wea, and Winnebago peoples. The name Chicago derives from the Miami word *sikaakwa* meaning "wild garlic," reflecting the seasonal harvest of fragrant, uncultivated ramps in the region. Along the North Branch of the Chicago River, in present-day LaBagh Woods Forest Preserve, researchers discovered a village site dating to the Archaic period and showing repeated use over thousands of years. These Indigenous peoples relied on the Chicago River for protein, collecting mussels from the muddy banks of the river, and netting fish and turtles from its waters. They hunted deer, bison, bears, beavers, otters, and muskrats along the river as well and collected maple syrup and nuts from hardwood trees in the area.[7]

Most importantly, Native American groups used their extensive knowledge of the geology and hydrology of the region to create trade networks connecting much of North America. According to Heather Miller, the former executive director of the American Indian Center in Chicago, "For time immemorial, tribes from across the country congregated here in the Chicago area; it was a central trade hub. Oftentimes we forget about that important history, and it's so often overlooked from the history of Chicago." At least seven thousand years ago, a vast trade network ran along the rivers and ridges of Chicago, allowing valuable material such as copper and vermillion from the Lake Superior region to move southward and Mississippian trade goods to flow north. Native Americans portaged between Chicagoland's two river systems in a variety of places, most notably along a short route known as Mud Lake. Native travelers could paddle from the South Branch through Mud Lake when it was wet enough, or tote their birch, basswood, or poplar canoes when it was dry, traveling about six miles to the Des Plaines River to the west.[8]

Europeans and eventually Americans came to appreciate the Pleistocene geology of Chicago. In 1673 Kaskaskia guides shared this invaluable information with Fr. Jacques Marquette and Louis Jolliet, Frenchmen searching the interior of North America for a northwest passage. Like Native Americans, Marquette and Jolliet envisioned a grand trade network connecting the Great Lakes region with the Mississippi River, and they correctly predicted that Chicagoland would be a key link in this commercial chain. On September 1, 1673, Marquette and Jolliet portaged through Mud Lake on their return trip to New France. Jolliet later reported that the portage was a prime location for a future canal. The Chicago River system had "a great and considerable advantage, and which one will maybe have a hard time believing: it is that we could easily go all the way to Florida by boat," Jolliet said. "There would only be one channel to make, cutting across one-half league of prairie to embark upon the bottom portion of the Lake of the Illinois [Lake Michigan] into the Saint-Louis River [Illinois River] . . . there the boat will be in position to easily navigate to the Gulf of Mexico."[9]

Over the next century, Native Americans and Europeans used the Chicago Portage as a nexus of trade—part of a region of multicultural diplomacy and "mutual invention" that historian Richard White refers to as the "middle ground." One such trader, a man of West African and French descent named Jean-Baptiste Point du Sable, became the first permanent non-Native settler in the Chicago region. In the late 1770s, du Sable built a trading post at the mouth of the Chicago River near the lake and married a Potawatomi woman named Kitihawa, a union that produced two children and cemented du Sable's position in Indigenous trade networks. While du Sable worked within the norms of the "middle ground," American arrivals prized the Chicago landscape and wanted to commodify the watery crossroads of America. Anglo-American squatters stole lands from local Indigenous peoples, pressing the federal government into war in Illinois. After a series of martial conflicts, which concluded with the defeat of a coalition of Sac, Fox, and Kickapoo led by Black Hawk in 1832, the U.S. government forced the signing of removal treaties in 1833.[10]

The Lockean conceptions of property held by the newest Chicagoans confused and angered Black Hawk and other Native American leaders. Black Hawk believed his people had a "right to the soil"—to benefit from the bounties of nature and the trade networks that crisscrossed the region—

but was adamant that "my reason teaches me that land cannot be sold. The Great Spirit gave it to his children to live upon and cultivate as far as necessary for their subsistence." In 1833, under pressure from federal leaders to sign treaties, Potawatomi leaders declared that "their Great Father in Washington must have seen a bad bird which had told him a lie, for that far from wishing to sell their land they wished to keep it." Recently, historian and Potawatomi John Low confirmed these conclusions, arguing that his people understand that "the land is Mother Earth. You can't own it—it's like owning air, owning the stars."[11]

The newest Chicagoans disagreed. In 1833, the same year local Native groups signed disputed treaties and moved west across the Mississippi River, local American leaders incorporated the village of Chicago and platted roads, sold lands in an orderly grid pattern, and made bold plans for the region's waterways. A new era had arrived.

William B. Ogden came to Chicago in 1835 as a land speculator and was elected the first mayor of Chicago two years later. Ogden developed into a giant in Chicago's history, a city advocate focused on the construction of needed infrastructure. Ogden pushed for the realization of Jolliet's prophecy of a canal linking the Great Lakes with the Mississippi River. Although Ogden was more interested in railroads, the canal idea fit his dream of a "visionary" and "splendid system" of transportation, linking his home state of New York with the booming areas of the Midwest, the Mississippi River, and eventually New Orleans. Newspaper editors, boosters, and other politicians also pushed for a Chicago canal as central to the growth of both the city and the nation. "Nature's design surely was not ambiguous in this instance—nor can a sagacious and enterprising people long neglect to complete what she has all but accomplished," one supporting newspaperman wrote in 1822. "To effect the junction of the St. Lawrence and the Mississippi were a task worthy of a great nation." That same year, the U.S. Congress agreed and appropriated funds and land grants for a canal connecting the Chicago and Illinois Rivers.[12]

Built between 1836 and 1848, the Illinois and Michigan Canal (I&M) ran ninety-six miles through the subcontinental divide, from Chicago to LaSalle, Illinois. As part of the construction, Ogden and his business partner received a $200,000 contract to build the first segment of the canal out of the South Branch of the Chicago River. On April 10, 1848, raucous crowds cheered as the *General Fry*—the first boat through the canal—entered Lake Michigan, and according to one reporter, "If there was a sober man in Chicago that night—*I did not see him!*"[13]

The construction of transportation routes through Chicago changed the course of history and allowed a metropolis to grow on the shores of Lake Michigan. In 1847 Ogden's political colleague Judge Jesse Thomas lamented that to date "Chicago has been merely the centre of a local retail trade of a few hundred miles of . . . country." Following the opening of the canal, agricultural goods moved through Chicago in record numbers. Corn shipments increased 800 percent the first year the canal operated, as Illinois farmers shifted their market allegiance from St. Louis to Chicago, and increased lumber sales moved south through the canal to build more downstate farms and towns. Historian William Cronon argues that the construction of the I&M Canal "was as if a corridor of relatively cheap transport had suddenly appeared like a fault across the . . . isolated city," thereby "necessitating a complex series of adjustments in the region's spatial economy." In 1905 alone almost seventy million bushels of cereal grains and more than two hundred thousand tons of milled flour moved through Chicago's waterways onto Great Lakes ships and eastward.[14]

More than the canal even, the first railroads through Chicago facilitated massive growth in the city. The first ten miles of the Ogden's Galena and Chicago Union Railroad opened the same year as the canal and connected Chicago to the Des Plaines River. Soon, the city became the railroad capital of the world. By 1893 an astonishing 4 percent of all railways on the planet terminated in Chicago, connecting thirty million people to markets, material goods, and geography unheard of only decades earlier. Combined, the linkage of the Great Lakes watershed with the Mississippi via the Chicago River and the veins of converging railroads modernized Chicago and turned a swampy town of a few hundred residents in 1833 into the fastest growing city in the world by 1900.[15]

The rise of industrial Chicago in the nineteenth century led directly to the environmental degradation of the Chicago River. In 1847 Cyrus McCormick relocated from Cincinnati to Chicago and opened the McCormick Harvesting Machine Company near the mouth of the Chicago River a year later. McCormick's three-story plant initiated the transformation of Chicago manufacturing from provincial to international in scope. By 1890 McCormick's company had sold half-a-million steel reapers, and Chicago was the epicenter of industry in the United States. Chicago's famed stockyards and slaughterhouses, led by innovators like Philip Armour and Gustavus Swift, produced more meat and animal byproducts than anywhere else in the country. Chicago led the nation in the production of lumber and furniture and the fabrication of cast-iron machinery and tools, and it was a leading producer of clothing, tobacco goods, raw iron, and steel. During the second half of the nineteenth century, Chicago truly remade itself into "the city that works," but at a significant environmental cost. At the dawn of the twentieth century, Robert Herrick described Chicago in his realist novel, *The Gospel of Freedom*. "Brazen, unequal, like all man's works," Herrick wrote, Chicago "stands [as] a stupendous piece of blasphemy against nature."[16]

From incorporation, Chicago leaders realized that building a city on a nearly stagnant swamp was not an ideal plan for maintaining a sanitary river. In 1833 city leaders issued an anti-pollution edict prohibiting the dumping of "dung, dead animal carrion, putrid meat, or fish entrails, or decayed vegetables, or any other offensive substance whatever . . . into the Chicago River." Violators faced an ineffective three-dollar fine. Industrial concerns and individual citizens alike unloaded their waste in the river. The most obvious culprits were the meatpacking plants along the South Branch, which dumped offal, blood, broken bones, unused fat, and other waste directly into the river. William McCormick wrote his brother Cyrus that "the river is positively red with blood under the Rush Street bridge and down past our factory! What a pestilence may result from it I don't know."[17]

Beginning in 1858, Chicago attempted to sanitize its built environment, which further polluted the river. Engineers raised the entire city out of the swamp and installed a modern sewer system underneath that dumped rainwater and human waste directly into the river. By the 1860s, the *Chicago Tribune* described a river composed of a "filthy compound exhaling a sickening odor . . . so intense that it permeates our dwellings and mingles its nastiness with the air we breathe, with the water we drink and the food we eat." The putrid river contaminated Chicago's drinking water and caused epidemic disease and death. A cholera outbreak in 1866, for example, killed almost one thousand Chicagoans, which Chicago's health officer linked to the improper disposal of offal and sewage, justifying the construction of a remarkable two-mile tunnel into Lake Michigan to secure a potable supply.[18]

The contamination of the Chicago River also killed wildlife in the watershed, most obviously fish populations. Before the rise of Chicago, numerous species of fish lived in the river. Native sturgeon, walleye, and pike collected in the deepest pools of the Chicago River, and brook trout spawned upstream in the spring. By the end of the nineteenth century, the river was nearly devoid of life. One observer claimed that when "fluid from the Chicago River" entered the Illinois River via the canal, "fish, unaccustomed to such an element, acted most strangely in springing out of the water, and some were instantly killed. The frogs all drew up in a line on the shore and began to croak [in distress and] . . . the action of the snakes was also peculiar, and it is thought the filth from the city named [Chicago] will drive fish, frogs, and snakes from the Illinois." When a fishmonger claimed to sell "Chicago River trout" in Baxter Springs, Kansas, the town's newspaper warned consumers that "there are no trout in the Chicago River; in fact, there is no more animal life in the Chicago River than there is in the Dead Sea." In 1892 strong east winds blew schools of minnows upstream into the river. According to one report, millions of minnows were quickly "smothered by the filthy water," and their floating, lifeless bodies "gave to the murky water an unwanted sheen."[19]

Outside observers proposed many solutions to the environmental ruin of the river. One newspaper editor from Rock Island, Illinois, argued that Chicago should end all pretense of wildness "since the stream is little better than a sewer, and ought to be covered over and used for that purpose." A Waukegan resident wanted to hold the industrial culprits responsible for the "evil" river. "To obviate this difficulty," the resident wrote, "I would in the first place declare the thousand and one distilleries, breweries, tanneries, slaughter houses, &c, a nuisance, and remove their slop and filth to some other locality remote from the city." Chicago, the fastest growing city in the entire world, was unwilling to banish its vast manufacturing base from the region and unable to police its river effectively.[20]

Instead, Chicago focused on various schemes to reverse the flow of the river and send its filth toward the Gulf of Mexico. Beginning in the 1860s, Chicago used hydraulic pumps to force contaminated water down the canal. According to one glib reporter, pumping stations in Bridgeport would help clean the fetid river and clarify Union efforts during the Civil War. "At the present rate of progress," the *Chicago Tribune* proclaimed, "all the filth in the river will, in a day or two, be on its way to the rebels. If the latter can stand it, it will be of no use to bombard them any longer."[21]

Chicago wanted a permanent reversal of the river and used clever engineering to accomplish this anthropocentric task. One plan involved deepening the existing I&M Canal, using gravity to pull lake water into the river to switch the flow, although this adjustment failed anytime heavy precipitation fell in the area. Between 1892 and 1900, Chicago built the Sanitary and Ship Canal, which connected the South Branch to the Des Plaines and reversed the river. When finished, the canal was thirty-four miles long, 290 feet at its widest section, and 22 feet at its deepest, with a projected flow capacity of 600,000 cubic feet per minute, making it three times bigger in terms of speed and volume than the I&M Canal. Some contemporary reports dripped with sarcasm—"Water in Chicago River Now Resembles Liquid" exclaimed the *New York Times*. City leaders, however, bragged to Congress that "it is no exaggeration to say that the Chicago Sanitary and Ship Canal is one of the greatest artificial waterways ever constructed," and the *Chicago Tribune* called it "one of the greatest engineering works of all time."[22]

The reversal of the Chicago River began the rewilding of the city's waterways, but continued human intervention impeded aquatic populations. The reversal certainly improved the water quality of the Main Branch nearest Lake Michigan, and the mouth of the river became a popular fishing spot, but most of the river remained a quagmire of pollution. That filth simply flowed in a different direction. To make the river safer, Chicago built a modern sewage system, with four massive treatment plants located throughout the Chicago Area Waterway System (CAWS). Human waste flowed into these facilities, where sanitation workers cleansed effluent before returning it to the Chicago River watershed. In 1955 the American Society of Civil Engineers named Chicago's sewage disposal system one of the "seven wonders of American engineering," alongside other feats like the Hoover Dam, the Empire State Building, and the Golden Gate Bridge. This modern system discharged water "pure enough to bathe in."[23]

This was true for humans, perhaps, but not for most kinds of indigenous fish. From 1916 to 1984, Chicago reclamation facilities effectively treated wastewater with chlorine to kill bacteria and other disease-causing pathogens. In the 1890s, the typhoid death rate in Chicago was sixty-seven per one hundred thousand people. With the reversal of the river and the use of chlorination, typhoid deaths fell to one per one hundred thousand Chicagoans in 1919—the lowest rate in the entire country. When released, however, this "pure" fluid bleached the Chicago River's water and prevented the return of all but the hardiest species of fish.[24] For at least one powerful politician, this was a problem that needed rectification.

Chicago's iconic mayor Richard J. Daley, who ran the city from 1955 until his death in late 1976, was an unlikely proponent of the rewilding of the Chicago River. Known as the "Boss," or to many Chicagoans as "hizzoner da mare," Daley was a Democratic machine politician focused on improving Chicago's reputation as "the city that works." As other Rust Belt towns like Cleveland and Detroit declined in the 1960s, Daley oversaw eye-catching growth requiring untold steel, concrete, and tarmac, including the construction of the Sears Tower (at the time the tallest building in the world), the major expansion of O'Hare Airport (the longtime busiest airport in the world), and one of the world's most complex expressway systems. Nationally, Daley is probably best remembered for his autocratic actions and inflammatory remarks during 1968's civil unrest. In the aftermath of riots following the assassination of Martin Luther King Jr., for example, Daley ordered police "to shoot to kill any arsonist . . . [and] shoot to maim or cripple anyone looting any stores in the city."[25]

Daley was also an avid fisherman with a soft spot for Chicago's neon river. According to Daley—a devout Catholic—a fishing boat was "the one place [on Earth] where you're closest to God." He liked to fish early in the morning along the lakeshore in search of non-native sport fish. Beginning in the late 1960s, Great Lakes states introduced millions of coho and Chinook salmon smolts into Lake Michigan as stock for sport fishing. After three-to-four-year lifespans, full-grown salmon returned to their release points on the lake's shorelines, including the Chicago-area harbors of Jackson Park, Diversey, and Waukegan, looking for suitable streams to spawn. In 1976 Daley landed a nineteen-pound Chinook near Chicago's McCormick Place, one of an estimated one hundred thousand Chinooks in Chicago waters that season, and gleefully told reporters that "Ma's going to bake it" with mayonnaise and stuffing.[26]

Salmon fishing was an economic engine and a source of civic pride for Daley. He once told entertainer Bob Hope that "we are now the greatest coho salmon fishing center in the nation" and bragged that Chicago was the "salmon capital of the world." His most outlandish goal was to see wild salmon spawning up the Chicago River. "People from the Loop could catch fish in the Chicago River and barbecue them on grills we'll put in lower Wacker Drive," Daley informed the House of Representatives' Subcommittee on Water Resources. "They can eat fish and have a bottle of beer."[27]

It was a ludicrous suggestion. For one, salmon instinctively seek currents and spawn upstream into the waterways of their birth. Following the reversal of the Chicago River in 1900, the river flowed away from Lake Michigan. Any stocked fish unfortunate enough to enter the backwards river would soon struggle in a turgid stream too slow and warm to support a salmon population. Most obviously, a century and a half of industrial pollution, human waste, and sewage treatment chemicals made Daley's dream an impossibility. In his article "Salmon for Hizzoner," *Sports Illustrated* columnist Rick Telander described the fishy scheme to a national audience and argued that the Chicago River was "an inner-city sludge pot of indeterminate composition and color . . . [that] had probably housed more cement-encased humans than fish during the last half century." He concluded that "the only time the river has a healthy color is on St. Patrick's Day, when the Democrats dye it green with food coloring."[28]

Daley's pipedream and Telander's sardonic response reflected two different visions of the Chicago River. Telander understood the nearly century-and-a-half history of the downtrodden aspects and industrialization of the neon river. The Chicago River was a product of anthropocentric views of nature. Chicagoans dredged canals to facilitate commerce, allowed almost unimaginable contamination in the river in support of industrial growth, and funded astonishing engineering efforts to keep a source of potable water for an exploding population. Each of these decisions benefited human beings, often at the expense of the ecological health of the river.

Daley envisioned the next half century of redemptive history of the Chicago River. Daley was not a staunch environmentalist, and his dream of a partially restored river was anthropocentric as well. With the passage of the Clean Water Act of 1972 and the subsequent rise in environmental concern, Chicago began the arduous process of cleaning up its river and

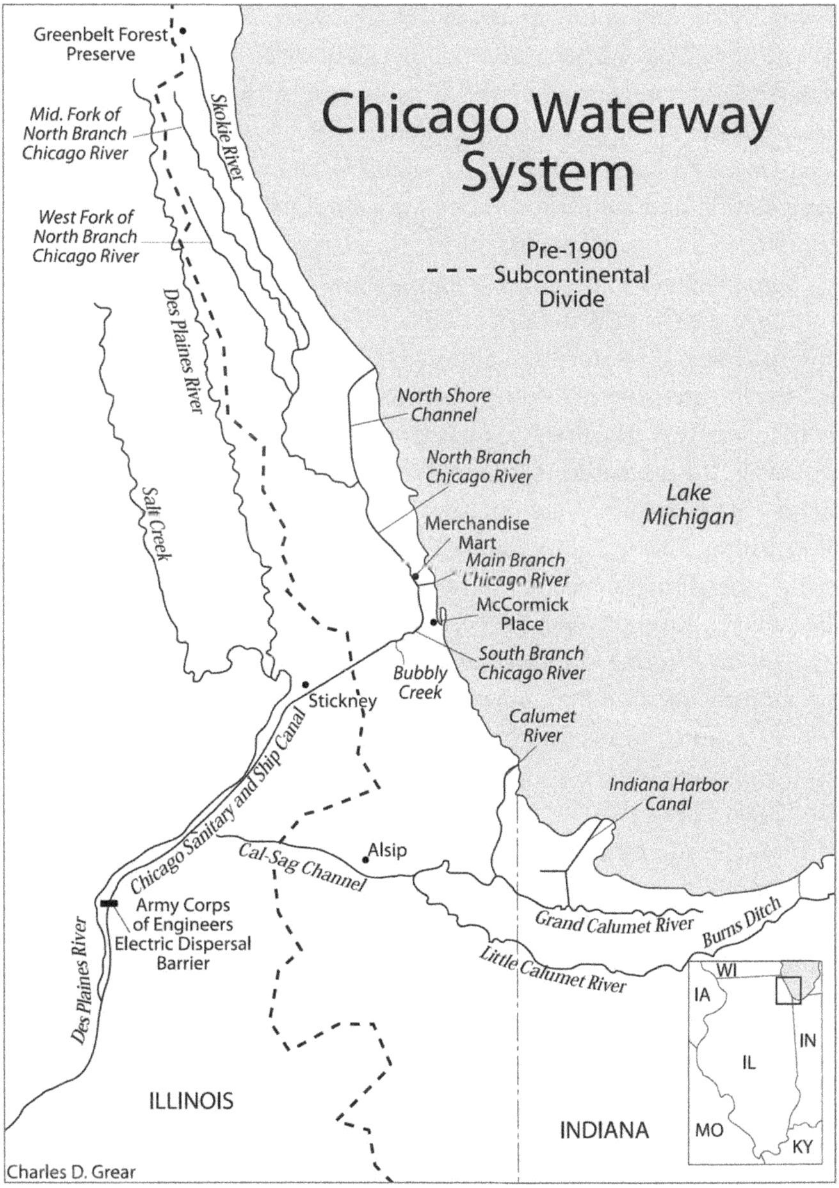

15. Map of Chicago waterways. Created by Charles David Grear, used by permission.

rewilding its waters for the benefit of Chicagoans. Although there are still no spawning salmon, today almost sixty species of fish swim in the watershed, and anglers along the Main Branch catch healthy largemouth bass, crappie, walleye, catfish, and other species in the shadow of skyscrapers. Combined, the history of Telander's "sludge pot of indeterminate composition and color" and Daley's dream of urban salmon reflect the humanized nature of Chicago during the onset of the Anthropocene and are both evidence of the formation of the neon river.

In 1972 the Clean Water Act established regulations concerning pollution in American waterways, although Chicago's compliance took time, and the maintenance of healthy ecosystems for fish and wildlife remains a work in progress. The first two objectives of the Clean Water Act included requiring "the discharge of pollutants into navigable waters be eliminated by 1985" and improving national "water quality . . . for the protection and propagation of fish, shellfish, and wildlife." In Chicago this meant (among many other things) ending the release of chlorine-treated effluent into the Chicago River. In 1984 the Metropolitan Water Reclamation District stopped discharging chlorine-treated water into CAWS, and the region's fish assemblage started to rebound.[29]

A 2021 study, which caught and identified almost fifty-four thousand fish in Chicago waters, concluded that there are now fifty-eight distinct species of fish in CAWS. While non-native (and pollution-tolerant) common carp remain the most abundant, improving populations of native shad, largemouth bass, bluegills, and other sunfish swim in Chicago's waters. According to biologist Austin Happel, the "actions we're taking are working. We're already seeing that if you can increase fish diversity, you're also able to increase how people use waterways."[30]

The improving ecological health of the Chicago River led directly to increased wildlife in the watershed. By the twenty-first century, river otters frolicked under elevated train bridges, beavers built dams at the base of Merchandise Mart, and white-tailed deer, red foxes, and coyotes traversed the city along the river's banks. In 2008 a full-grown cougar migrated into the Roscoe Village neighborhood on the city's northside, likely by following parts of the Chicago River system.[31]

Like the European common carp, not all species in the Chicago River are native to the region, contributing to the waterway's reputation as an "invasive species superhighway." An estimated 180 invasive species live in the Great Lakes and 163 are in the Mississippi watershed. According to the Army Corps of Engineers, about 40 of these non-natives might jump ecosystems via this arterial expressway. For example, hull-hugging zebra and quagga mussels, which arrived in the ballast water of European ships, are migrating from Lake Michigan into the American West, where they have the potential to wreck hydroelectric dams on the Missouri and Columbia Rivers. According to one estimate, non-native mussels would cause $250 million in annual damages on the Columbia. Other aquatic threats include the round goby, rusty crayfish, northern snakehead, and spiny water flea.[32]

Invasive species can also go the other way. First introduced by southern catfish farmers in the 1970s, Asian carp are voracious eaters with twenty-year lifespans and the ability to dominate ecosystems. After Asian carp escaped holding ponds and flourished in the Mississippi watershed, they entered the Illinois River and continued migrating northeast toward Lake Michigan. The Army Corps of Engineers installed electric fish barriers near Romeoville, Illinois, to halt this progression in the Sanitary and Ship Canal and keep the non-native fish out of the Great Lakes.[33]

The reconfigured Chicago River is a river of the Anthropocene. Popularized by Nobel Prize-winning chemist Paul Crutzen in 2000, the Anthropocene is the notion that Earth has entered a new geologic age marked by the domination of human beings. According to Crutzen and his colleague Eugene Stoermer, massive population growth and urbanization, the near-exhaustion of fossil fuels, the use of half of all the planet's freshwater, and the physical transformation of upwards of 50 percent of lands on Earth are all evidence of this new geologic era. They proposed a start date of the late 1700s, reflecting James Watt's invention of the steam engine and the onset of the Industrial Revolution. More recently, many scientists have proposed 1952 and the entrenchment of the Atomic Age as the beginning of the Anthropocene. Given the scope of geologic time—Earth is, after all, at least 4.54 billion years old—a very specific origin point seems capricious in nature. The future is much more important. As Crutzen and Stoermer put it, "Without major catastrophes . . . mankind will remain a major geological force for many millennia, maybe millions of years to come. To develop a worldwide strategy leading to sustainability of ecosystems against human induced stresses will be one of the great future tasks of mankind."[34]

Today, the Chicago River is a humanized landscape. According to the Illinois Environmental Protection Agency, CAWS contains seventy-eight miles of canals and modified rivers, including most of the Chicago and Calumet watersheds. Humans dominate every mile of this system, which is 75 percent man-made canals and 25 percent "natural streams" that have been dredged, dammed, concreted, straightened, and broadened for the benefit of Chicagoans. The main purpose of CAWS has not changed much in the past two centuries, as Chicago's waterways "provide an outlet for drainage of urban storm water runoff," including "treated municipal wastewater effluent," as well as "support commercial navigation." According to the Army Corps of Engineers, 70 percent of the annual water flowing out of CAWS is cleansed wastewater. Nonhuman nature is secondary in this system.[35] The Chicago River, and all its built and modified components, is truly a river of the Anthropocene. And as Crutzen and Stoermer suggested, Chicagoans must understand the complex history of their neon river as they imagine their future.

Daniel Burnham wanted Progressive Era planners to think big. In 1910 Burnham—Chicago's most famous architect and urban developer known for his skyscrapers, his management of the 1893 World's Columbian Exposition, and his 1909 *Plan of Chicago*—gave a speech in London, England, connecting twentieth-century urbanity with the future of American democracy. In one of the most quoted phrases in Chicago history, Burnham exclaimed, "Make no little plans; they have no magic to stir men's blood and probably themselves will not be realized. Make big plans; aim high in hope and work . . . [and] remember that our sons and grandsons are going to do things that stagger us."[36]

Less well-known are the environmental proposals of Burnham's speech. Chicago would be "the city of the future," he predicted, a modern, sanitary environment "that would make our forebears think us magicians." According to Burnham, "Our city of the future will be without smoke, dust or gasses from manufacturing plants, and the air will therefore be pure. The streets will be as clean as our drawing rooms today." Water played a vital role in this projection, as citizens must conserve potable supplies while maintaining recreational access to the lakefront and riverbanks. "Do [all of] this because of the effect of nature upon citizenship," Burnham concluded. “Other things being equal, a person accustomed to living in nature has a distinct advantage all his life over the purely town-bred man. Allure your city denizen to sylvan nature, for it is there he finds the balm his spirit needs.”[37]

Throughout its history, the “City of the Big Shoulders” was also the city of the “big plans.”[38] The first Chicagoans organized trade networks that connected much of the continent through the Chicago River portage. Marquette and Jolliet used this Indigenous knowledge to envision a French empire linked by water, from Quebec to Florida. In the nineteenth century, Chicagoans brought this concept to fruition, building canals that linked the Atlantic Ocean with the American West. As Chicago grew into an industrial giant, the Chicago River morphed into an open sewer, which threatened the city’s drinking water. To protect the health of the city, Chicago proposed the biggest scheme of all—the reversal of the Chicago River. Chicago built a “wonder” of a sewage-treatment system, and in the 1970s, Mayor Daley reimagined the river as a salmon-spawning ground. This daring needs to continue. As Chicago looks to the future, it should embrace the audaciousness of this history and both aspects of Burnham’s “city of the future,” making no little plans to ensure the environmental health of its river.

Historically, Chicago leaders used Promethean philosophy in their “big plans” to alter the river. In Greek mythology, a Titan named Prometheus stole fire from Zeus and passed this life-altering knowledge to mortals. In essence, Prometheus gave the Greeks technology and the basis for civilization. Political scientist John Dryzek uses this myth to explain human optimism toward ecological issues. According to Dryzek, the central tenets of the “Promethean response” to industrialism include the “denial of environmental limits” and an “unlimited confidence in the ability of humans and their technologies to overcome any problems—including environmental problems.”[39] Ultimately, Prometheans embrace rather than fear the Anthropocene. Human domination of Earth’s geology and natural processes is a feature in this new era, not a bug.

This has been the Chicago standard mindset regarding the environment throughout most of its history—to implement technological solutions to human-caused environmental problems. Contamination, disease, stench, invasive species—these have not been reasons to curb the abuse of the

Chicago River, but rather these are the justifiable results of human progress and opportunities for greater growth. Through godlike machinations, Chicago's modern-day Titans can continue to mitigate the deprivation of the neon river in perpetuity, at least theoretically.

Since 1972 one of Chicago's most Promethean projects has been the Tunnel and Reservoir Plan (TARP)—colloquially known as "Deep Tunnel"—of the Metropolitan Water Reclamation District of Greater Chicago (MWRDGC). In a continued effort to clean up the Chicago River and meet the requirements of the Clean Water Act of 1972, Chicago city and suburban engineers designed an underground river system longer than the Chicago River itself. Chicagoland's antiquated sewer systems collected human waste and rainwater in the same pipelines, throughout 375 square miles consisting of the city of Chicago and fifty-one separate suburbs. During dry conditions, the existing system diverted this water through treatment plants, which removed effluent before discharging water into the Chicago River and other area watersheds. During periods of moderate rainfall, however, the system became overburdened. Prior to the completion of TARP, one-third of an inch of rainfall was enough to overload Chicago's combined sewer system, and human waste dumped into the Chicago River about one hundred days every year.[40]

Civil engineers designed TARP to alleviate this problem and broke ground on phase one of the estimated $1.2 billion project in 1975. TARP is a subterranean, man-made river system that spiderwebs more than 110 miles throughout Chicagoland. Beginning in 1981, concrete tunnels, sometimes thirty-three feet in diameter and buried more than three hundred feet below grade, intercepted excess water from the congested sewer systems and stored this liquid underground. When rains and potential floodwaters abated, reclamation engineers rerouted the water into treatment facilities and eventually back into the Chicago River. This improved system was not infallible, however. On October 3, 1986, for example, rainwater overwhelmed the partially completed TARP, causing wave action within the tunnels and at least three impromptu geysers of rainwater and human excrement. At 8:00 a.m. a woman named Dolores Crain stopped her Pontiac at a red light—unknowingly over a TARP tunnel grate—in Chicago's Loop while on her way to work. Pressurized water from the sewer system shot thirty feet into the air and lifted Crain's Bonneville up at a forty-five-degree angle “like a toy.” The filthy geyser lasted about two minutes, leaving Crain a drenched and disgusted “bundle of nerves.”[41]

Work on the parallel river system continued for another twenty years, with mixed success. Completed in 2006, TARP’s concrete tunnels have a maximum capacity of 2.4 billion gallons of water. Phase two of TARP began in 1990 and included the construction of three massive reservoirs scattered throughout the area. When it will be completed in 2029, the fully operational TARP will have a maximum storage potential of 17.5 billion gallons of water. That still might not be enough to protect Chicago from flooding and prevent raw sewage from flowing into the river. From 2007 to 2011, despite the completion of the underground tunnels, the MWRDGC dumped almost 19 billion gallons of contaminated water in the Chicago River. Environmental analysts point to the increased intensity and frequency of storms in Chicago, most likely due to human-induced climate change, and argue that the now more than $4 billion TARP system may be obsolete before its even finished. According to Karen Hobbs, Promethean optimism was misguided regarding TARP and the Chicago River watershed. A former deputy environmental commissioner for the city of Chicago, Hobbs argues that TARP is “a marvel. But we have this tendency in this country to think we can build our way out of stuff. And we can’t always build our way out.”[42]

The Chicago standard might also provide a solution to the problem of invasive species in the region’s waterways. To prevent the spread of “aquatic nuisance species” between the Great Lakes and Mississippi River watersheds, many leaders advocate technological solutions resulting in the return of the subcontinental divide in Chicago and the re-reversal of the Chicago River. In 2012 the Great Lakes Commission released a $2 million study that called for such a result. By installing man-made barriers at key points along the subcontinental divide, Chicago could end the artificial connection between its two distinct watersheds and prevent the migration of Asian carp into the Great Lakes. According to David Ulrich, a former deputy regional administrator of the U.S. Environmental Protection Agency and chief architect of the study, “If 110 years ago, from an engineering and financing standpoint, we could dig this huge canal and reverse (the river), 110 years later, with all the advances in technology, we certainly could undo what we have done.” This undoing would cost around $4 billion, while allowing Chicago's canal-shipping industry to continue to thrive. Sixteen million tons of cargo move through the Sanitary and Ship Canal annually, an economic reality that all sides want to maintain. Initially, the Chicago River would continue to flow southwestward. After the completion of TARP in 2029, combined with proposed improvements to water-treatment protocols in Chicago, the river would once again permanently flow into Lake Michigan for the first time since 1900.[43]

The Army Corps of Engineers studied the potential movement of invasive species through Chicago's watersheds and released a report of potential solutions called the Great Lakes and Mississippi River Interbasin Study (GLMRIS) in 2014. Like the Great Lakes Commission, the Corps considered the restoration of the subcontinental divide and the re-reversal of the Chicago River as one potential solution, a plan they termed "Mid-System Hydrologic Separation." In this proposal, the Army Corps of Engineers would build two physical barriers and two new water treatment plants in Chicagoland—one in Stickney, Illinois, to stop the Chicago River and one in Alsip, Illinois, to block the Calumet River system. These barriers would harm the rebounding indigenous fish populations nearest the barriers due to water stagnation but with the benefit of "ultimately reversing an environmental impact from previous human activities."[44]

If completed, the Chicago and Calumet Rivers would once again flow into Lake Michigan but so would Chicago's drain water and sewage overflow. This could be mitigated, the Corps argued, with advanced but high energy consuming water treatment techniques usually implemented in water-scarce locations. The entire plan would be expensive. According to the Army Corps of Engineers, the re-creation of the divide and the re-reversal of the river would cost more than $15.5 billion and take twenty-five years to complete.[45] Due to global climate change during the Anthropocene, the re-reversal might not even work, at least not permanently.

Between 2013 and 2020, water levels in Lake Michigan fluctuated at an astounding rate. In 2013 lake levels plunged to a record low due to high evaporation, threatening drinking water supplies in the city of Chicago. The next year the lake began rising, increasing in height six feet over the next seven years. Many scientists attribute these historic swings to human-induced climate change. When Great Lakes winters are mild and summers severe, intensified evaporation causes critical drops in lake levels throughout the region. Since 1991 the average air temperature in the Great Lakes region has increased 1.9 degrees Fahrenheit, with no signs of abating. If Lake Michigan drops low enough, the Chicago River might re-reverse on its own.[46]

Climate change can explain rising lake levels as well. Beginning in 2014, Chicagoland suffered through intensely cold winters due to the dislocation of the Polar Vortex, likely initiated by global warming. In January 2019, for example, the state of Illinois set a record low of -38°F west of Chicago, Lake Michigan ice levels increased, and evaporation rates plummeted during these extreme winters. When temperatures rebounded in spring and summer, the ice melted and Lake Michigan rose to historic heights, destroying beaches, flooding streets, and threatening lakefront high-rises in Chicago. Increasingly intense storms also led to rising lake levels. Historically, Chicago receives an average of 4.49 inches of rainfall each May. In May 2018 Chicago set a record with 8.21 inches of rain. The next year 8.25 inches of rain fell on Chicago in May. In May 2020 a record-shattering 9.51 inches drenched the Windy City. Due to all these factors—lower evaporation rates, the displacement of the Polar Vortex, and the increasing intensity of storms—some models predict an additional two-foot rise in Lake Michigan water levels.[47] If that happens, the $15.5 billion dream of re-reversing the Chicago River might be moot, since the lake would be significantly higher than the river.

Combined, the TARP system and the proposed re-reversal of the Chicago River are modern efforts to improve Chicagoland watersheds that match the Chicago standard throughout the history of the neon river. Through ingenuity, technical expertise, and no small amount of hubris, Chicago has attempted to overcome environmental degradation of the river's past. The goals are worthy—to control sewage contamination, to mitigate flooding, and to improve the overall health of the Chicago River. The question is what should happen in the future? Is this all Chicago could be doing to protect the river? If completely implemented, these joint efforts will cost almost $20 billion with no guarantee of long-term success. Both plans underestimated or misunderstood the complex hydrological changes wrought by human-induced climate change, and both ignored the realities of the Anthropocene. Perhaps it is time for the city to take an additional step in its environmental story and reimagine the concept of the Chicago River itself. In doing so, Chicago can add a strong legal protection to its vital waterways as it works to mitigate issues of the Anthropocene.

One bold protection would be granting the rights of personhood to the Chicago River—a philosophical idea with Indigenous roots. In the early 1970s, Walt Disney Productions wanted to build a thirteen-thousand-acre ski resort in the Mineral King Valley of California's Sierra Nevada, within Sequoia National Game Refuge adjacent to Sequoia National Park. In response, the Sierra Club sued to prevent the construction of the resort. In 1972, as the case worked its way to the Supreme Court, law professor Christopher Stone published "Should Trees Have Standing?" to argue for legal rights for environmental entities. Stone declared, "I am proposing we do the same with eagles and wilderness areas as we do with copyrighted works, patented inventions, and privacy: *make* the violation of rights in them to be a cost by declaring the 'pirating' of them to be an invasion of a property interest." In essence, humans cannot own the waters of a river since they did not create them—an echo of the beliefs of many Indigenous Chicagoland cultures.[48]

Instead, Stone wanted the valley itself to have standing as the plaintiff, not environmental groups like the Sierra Club with only "tenuous" connections to the ecological damage. Stone argued that "if the courts could be persuaded to think about the park as a jural person—the way corporations are 'persons'—the notion of nature having rights would here make a significant operational difference." In this theory, guardians of the environment could speak for environmental entities in the court system and argue for the protection of a river's rights, but the river itself would be the legal plaintiff.[49]

Supreme Court Justice William O. Douglas—the "Environmental Justice" according to one biographer—agreed in his famous dissent in *Sierra Club v. Morton* in 1972. A 4–3 majority ruled that the Sierra Club did not have standing to sue in court because the development of a remote portion of Sequoia National Forest by Disney would not materially injure the Sierra Club. In his dissent, Douglas concurred with Stone and declared that "the river as plaintiff speaks for the ecological unit of life that is part of it. Those people who have a meaningful relation to that body of water—whether it be a fisherman, canoeist, a zoologist, or a logger—must be able to speak for the values which the river represents, and are threatened with destruction."[50]

While it was a fringe environmental legal philosophy in the 1970s, the concepts of rights and personhood for nature have emerged in the twenty-first century as a viable legal protection for worldwide rivers in the Anthropocene. In 2017 New Zealand designated the Whanganui River to have all the "rights, powers, duties, and liabilities of a legal person" and declared the river a spiritual ancestor of the Indigenous Whanganui Iwi peoples. Two years later, Bangladesh granted personhood status to all its rivers. According to one Bangladeshi environmentalist, "The river is now considered by law, by code, a living entity, so you'll have to face the consequence by law if you do anything that kills the river."[51]

In North America, numerous localities have followed suit. In 2017 an environmental group called Deep Green Resistance sued the state of Colorado in an attempt to gain personhood rights for the Colorado River, an effort that failed when Colorado threatened to sanction the filing attorney. In 2019 the Yurok Tribe of Northern California established personhood for the Klamath River. For the Yurok, the Klamath is not a resource to use or exploit, but rather an entity with its own right to exist, especially when threatened by climate change during the Anthropocene. Also in 2019 the citizens of Toledo, Ohio, passed a referendum called the "Lake Erie Bill of Rights," which argued that "Lake Erie, and the Lake Erie Watershed, possess the right to exist, flourish, and naturally evolve." A federal judge later struck down the Toledo ordinance as vague and unenforceable because Toledo did not have jurisdiction over all of Lake Erie. In 2020 the Menominee Tribe of Wisconsin passed a resolution recognizing the rights of the Menominee River, which "possesses inherent and legal rights including the right to naturally exist, flourish, regenerate, and evolve; the right to restoration, recovery, and preservation . . . and the right to be free of activities and practices, as well as obstructions, that interfere with or infringe upon these rights." In 2021 two separate municipalities granted personhood to the Magpie River in Quebec, the first such designation in Canada.[52]

For Maia Wikaira, a descendent of the Ngāti Tūwharetoa, Te Rarawa, and Ngāpuhi Maori of New Zealand and a lawyer who aided the Yurok Tribe's legal efforts to protect the Klamath River, these modern efforts reflect the beliefs of Indigenous groups throughout the world. According to Wikaira, "It's another example of where long-held Indigenous perspectives and association with the natural world are not only being embedded within our legal system—they're being seen in popular environmental movements as an innovative way forward and a necessary step. So, old is new again."[53]

How would personhood work for the Chicago River? With designated rights and personhood, the river would shift from an object of the law to a subject. No longer would the waters, plants, fish, and animals be considered property to argue over, but instead the Chicago River would be an entity with legal standing. The Chicago River would have the intrinsic right to exist, regardless of its usefulness to human beings. While this might not undo past transgressions, personhood would strengthen the environmental protection of the river since the injured party moving forward would be the river itself. In *Sierra Club v. Morton*, the Supreme Court ruled that the Sierra Club had no standing in court since neither the club nor its property would be injured by the development of a ski lodge; with personhood, nature can stand for itself.

This would not be an easy transfer of rights. Detractors might balk at the granting of rights to nonhuman entities. If corporations are considered people in the U.S. legal system, however, why not rivers? As Christopher Stone pointed out, Americans in power have loathed granting rights to others. At various moments throughout U.S. history, it was "frightening or laughable" to the powerful to protect the liberties of the "rightless," including women, children, African Americans, Native Americans, and the mentally ill, as well as nonliving entities such as trusts and corporations. This discomfort, however, should not justify the denial of rights to the river. "There will be resistance to giving the thing [nature] 'rights' until it can be seen and valued for itself," Stone concluded. "Yet it is hard to see it and value it for itself until we can bring ourselves to give it 'rights'—which is almost inevitably going to sound inconceivable to a large group of people."[54]

The biggest hurdle would be bureaucratic jurisdiction. The city of Chicago could declare personhood for its river—a philosophical move that would demonstrate a commitment to the long-term health of its watershed. This would require the city to adjust its current development plan for the river. "The Chicago River is one of the city's most treasured and accessible natural resource," the city contends, and needs "a vision and set of standards for new development that will increase public access and create new recreational opportunities for all the city's residents." With personhood for its river, Chicago would most likely face the same legal challenges as Toledo and its Lake Erie Bill of Rights, since the state of Illinois "owns" the water and environment of the Chicago River. Therefore, the state would have to grant personhood to the river. State law requires the government "at all times to exercise a vigilant care to see that none of said bodies of water are encroached upon, or wrongfully seized or used by any private interest in any way, except as may be provided by law." In this legal existence, all rivers, lakes, and streams in Illinois are currently the property of the state. The federal government would still have some say as well, since federal pollution laws would be enforceable in Chicago, and the Army Corps of Engineers controls access to all navigable waters in the country. While the Corps states that its "strategy for watershed studies is to work in partnership with other interests on providing a shared vision with a holistic focus on water resource challenges," the overall purpose of the Corps focuses more on civil works such as harbor maintenance and hydroelectric power generation.[55] The city of Chicago, the state of Illinois, and the Army Corps of Engineers would all have to relinquish some control or goals in order for personhood to work for the Chicago River. None of these three entities is famous for its institutional elegance or rapidity of change. Bureaucratic lethargy, however, is not a valid reason to ignore the possibility of personhood for the Chicago River.

Nelson Algren once wrote that Chicago was "the place built out of Man's ceaseless failure to overcome himself. Out of Man's endless war against himself we build our successes as well as our failures. Making it the city of all cities most like Man himself—loneliest creation of all this very old poor earth."[56] This is the great lesson of the Anthropocene. Human beings must overcome themselves for the benefit of all. While we are not alone on this planet, we are lonely. Human beings are the only species that make conscious decisions regarding the environment, and our choices altered the Earth enough that we created a new geological epoch. Yet, we are also the one species with the ability to rectify mistakes through deliberate acts. As Chicagoans reconcile the long, complex, and often destructive history of their river and face the unknown millennia of the Anthropocene, the granting of personhood and legal rights would be a bold step toward protecting the Chicagoland environment. By making the neon river more like us in a legal sense, we can accelerate this necessary process.

NOTES

1. Algren, *The Neon Wilderness*, 3; Algren, *Chicago*, 11.
2. Doyle, *The Source*, 179; Dan Egan, "Deep Trouble," part 1, *Milwaukee Journal Sentinel*, December 15, 2012.
3. Katz, *Nature as Subject*, 94–96; Rosenzweig, *Win-Win Ecology*, 1–2.
4. Hill, *The Chicago River*, 16–17.
5. Hill, *The Chicago River*, 23–24; Chrzastowski, "Chicagoland," 32.
6. Kathy Paur, "Mammoth Bones Challenge History," *Chicago Tribune*, July 21, 2002.
7. Scarborough et al., *Adversity and Resiliency*, 7; McCafferty, "A Fresh Look," 116–17; Hill, *The Chicago River*, 34–35.
8. Heather Miller quoted in Daniel Hautzinger, "'We're Still Here': Chicago's Native American Community," WTTW, November 8, 2018, https://interactive.wttw.com/playlist/2018/11/08/native-americans-chicago; Rose and Associates, *The Chicago Portage*, 9; Hill, *The Chicago River*, 36.
9. Winslow, *Indians of the Chicago Region*, 29; Louis Jolliet quote in Margry, *Découvertes et éstablissements des français*, 267–68.
10. White, *The Middle Ground*, 50; Gustaitis, *Chicago in 50 Objects*, 17; Cronon, *Nature's Metropolis*, 27–29; Thwaites, *The Story of the Black Hawk War*, 49–50; Jung, *The Black Hawk War of 1832*, 50–53; Gerstle, *Liberty and Coercion*, 51.
11. Black Hawk, *Black Hawk's Autobiography*, 84; Latrobe, *The Rambler in North America*, 151; Jesse Dukes, "Without Native Americans, Would We Have Chicago as We Know It?," WBEZ, November 12, 2017, https://interactive.wbez.org/curiouscity/chicago-native-americans/.
12. William B. Ogden, "Speech at Albany, 1835, on the Erie Railway Bill," quoted in Arnold, *William B. Ogden*, 7; "The Western Lakes," *Long-Island Star*, March 7, 1822; "Navigation," *New York Daily Herald*, April 19, 1848.
13. Harpster, *The Railroad Tycoon*, 68; "Scenes in the Far-Off West," *Brooklyn Evening Star*, April 12, 1848.
14. Thomas, "Statistics Concerning the City of Chicago," 191; Cronon, *Nature's Metropolis*, 64–65; Sanitary District of Chicago, *Report on the Sanitary and Ship Canal*, 16.
15. Cronon, *Nature's Metropolis*, 65–67, 92–93; Ralph, *Our Great West*, 12.
16. Coclanis, "Business of Chicago," 42–44; Herrick, *The Gospel of Freedom*, 103–4.
17. "Laws and Ordinances of the Town of Chicago," *Chicago Democrat*, August 19, 1835; Hill, *The Chicago River*, 79–85; William McCormick quoted in Bluestone, *Constructing Chicago*, 207.
18. "The Water Question," *Chicago Tribune*, July 26, 1862; "Pure Water for Chicago," *Chicago Tribune*, March 26, 1867; "Health of the City," *Chicago Tribune*, April 3, 1867.

19. Hill, *The Chicago River*, 32; "River Intelligence," *Public Ledger* (Memphis TN), August 8, 1871; "City and County News," *Baxter Springs News* (Baxter Springs KS), February 6, 1892; "Chicago River Kills Fish," *Green Bay Press-Gazette* (Green Bay WI), February 16, 1898.
20. "Chicago Stenches," *Rock Island Argus* (Rock Island IL), September 1, 1862; J. A. F., "Stench of Chicago," *Waukegan Weekly Gazette* (Waukegan IL), June 13, 1863.
21. "Glorious News from Bridgeport—The River Filth Going to the Rebels," *Chicago Tribune*, May 30, 1862.
22. "Off for the Gulf," *Chicago Tribune*, July 16, 1871; Sanitary District of Chicago, *Proceedings of the Board of Trustees*, 6370–72; "Water in Chicago River," *New York Times*, January 14, 1900; "Chicago's Quest for Pure Water," *Chicago Tribune*, January 2, 1900.
23. Wolfert, "The Seven Wonders of American Engineering," 123, 128.
24. Cain, *Sanitation Strategy*, 127; Michael Hawthorne, "Nearly 60 Different Types of Fish Found in Chicago Waterways, Study Shows. 30 Years Ago, There Were Fewer Than 10," *Chicago Tribune*, August 3, 2020.
25. Bob Warden, "Mayor Daley's Legacy," *Washington Post*, January 1, 1977; Edward Schreiber, "Shoot Arsonists: Daley," *Chicago Tribune*, April 16, 1968.
26. Jerome Holtzmann, "The Greatest Thrill . . . Is to Take Your Son Fishing," *Chicago Sun Times*, December 21, 1976; Parsons, *History of Salmon in the Great Lakes*, 24, 49–50; Casey Bukro, "There's Another Angle to Fishing in the Chicago River," *Chicago Tribune*, November 9, 1978; "Daley's Catch," *Chicago Sun-Times*, September 18, 1976; Harry Golden Jr., "They Fall for Daley: Hook, Line and Sinker," *Chicago Sun-Times*, September 18, 1976; "Daley Lands 19-Pound Salmon," *Orlando Sentinel* (FL), September 19, 1976.
27. Casey Bukro, "Mayor Daley's Angling Ways May Make Chicago a Fishy Place," *Chicago Tribune*, November 4, 1976; Irv Kupcinet, "Kup's Column," *Chicago Sun-Times*, December 22, 1976; Tom McNally, "Chicago River Sport Fishing 'Only a Dream,'" *Chicago Tribune*, April 14, 1973; Telander, "Salmon for Hizzoner," 60.
28. Telander, "Salmon for Hizzoner," 60.
29. U.S. Congress, *An Act to Amend the Federal Water Pollution Control Act*, Pub. L. No. 92-500 (1972); Hawthorne, "Nearly 60 Different Types of Fish Found in Chicago Waterways."
30. Happel and Gallagher, "Chicago's Fish Assemblage," 313–15.
31. "Chicago's Wild Tale: How Fish and Birds Returned," *Chicago Tribune*, August 9, 2020; Nara Schoenberg, "Urban Otters in Spotlight at Wild Things Conference," *Chicago Tribune*, February 12, 2017; Noreen S. Ahmed-Ullah, "I'll Be Dammed! Wildlife Moves to Chicago River," *Chicago Tribune*, August 5, 2005; Jeremy Manier and Tina Shah, "Cops Kill Cougar on North Side," *Chicago Tribune*, April 15, 2008.

32. Egan, "Deep Trouble," part 1; Andrew Stern and Sandra Maler, "Plans to Block Carp Will Re-Reverse the Chicago River," Reuters, January 31, 2012, https://www.reuters.com/article/us-usa-greatlakes-carp/plans-to-block-carp-will-re-reverse-chicago-river-idustre80u1xl20120131.
33. Egan, "Deep Trouble," part 1; Stern and Maler, "Plans to Block Carp"; Carson Vaughan, "Floods, Carp, and Crap," NPR, October 14, 2019, https://www.wbez.org/stories/floods-carp-and-crap-the-environmental-impacts-of-the-chicago-river-reversal/df408b2a-16f0-44e1-8686-2668163993d2.
34. Crutzen and Stoermer, "The Anthropocene," 17–18; Rebecca Hersher, "Drawing a Line in the Mud: Scientists Debate When 'Age of Humans' Began," NPR, March 17, 2021, https://www.npr.org/2021/03/17/974774461/drawing-a-line-in-the-mud-scientists-debate-when-age-of-humans-began.
35. "Chicago Area Waterway System Use Attainability Analysis" (Springfield: Illinois Environmental Protection Agency, November 2004), section 1:4–6, https://www.csu.edu/cerc/documents/ChicagoAreaWaterwaySystemUseAttainabilityAnalysisUAA-November2004.pdf; "Summary of GLMRIS Report" (Washington DC: U.S. Army Corps of Engineers, January 6, 2014), 4, https://glmris.anl.gov/documents/docs/glmrisreport/GLMRISSummaryReport.pdf.
36. "Stirred by Burnham, Democracy Champion," *Chicago Record-Herald*, October 15, 1910.
37. "Stirred by Burnham, Democracy Champion."
38. Sandburg, "Chicago," 3.
39. Dryzek, *The Politics of the Earth*, 53.
40. "Tunnel and Reservoir Plan (TARP)," Metropolitan Water Reclamation District of Greater Chicago, accessed December 27, 2021, https://mwrd.org/tunnel-and-reservoir-plan-tarp; Henry Grabar, "Tunnel Vision," *Slate*, January 2, 2019, https://slate.com/business/2019/01/chicagos-deep-tunnel-is-it-the-solution-to-urban-flooding-or-a-cautionary-tale.html.
41. Flood Control Coordinating Committee, *The Chicago Underflow Plan*, 23; "Tunnel and Reservoir Plan (TARP)"; Grabar, "Tunnel Vision"; Rob Karwath, "Tunnel's Geyser Effect Still Puzzling," *Chicago Tribune*, October 17, 1986.
42. "Tunnel and Reservoir Plan (TARP)"; Grabar, "Tunnel Vision," including Hobbs quote.
43. "Summary of GLMRIS Report," 1; Egan, "Deep Trouble," part 1; Egan, "Deep Trouble," part 2, *Milwaukee Journal Sentinel*, December 17, 2012.
44. "Summary of GLMRIS Report," 150–62.
45. "Summary of GLMRIS Report," 150–62.
46. Dan Egan, "A Battle between a Great City and a Great Lake," *New York Times*, July 17, 2021, https://www.nytimes.com/interactive/2021/07/07/climate/chicago-river-lake-michigan.html.

47. Egan, "A Battle"; Elvia Malagon, "Polar Vortex Broke State Record," *Chicago Tribune*, March 11, 2019.
48. Ferguson and Bryson, "Mineral King," 493–95; Stone, "Should Trees Have Standing?," 450, 476.
49. Stone, *Should Trees Have Standing?* xiii.
50. Sowards, *The Environmental Justice*, 8; Sierra Club v. Morton, Sec. Int., No. 70-34, 405 U.S. 727 (April 19, 1972).
51. Ashley Westerman, "Should Rivers Have Same Legal Rights as Humans?," NPR, August 3, 2019, https://www.npr.org/2019/08/03/740604142/should-rivers-have-same-legal-rights-as-humans-a-growing-number-of-voices-say-ye; Kramm, "When a River Becomes a Person," 307.
52. Julie Turkewitz, "Corporations Have Rights. Why Shouldn't Rivers?," *New York Times*, September 26, 2017; Anna V. Smith, "The Klamath River Now Has the Legal Rights of a Person," *High Country News* (Paonia CO), September 24, 2019, https://www.hcn.org/issues/51.18/tribal-affairs-the-klamath-river-now-has-the-legal-rights-of-a-person; "Ohio EPA May Finally Clean Up Lake Erie," *Akron Beacon Journal* (Akron OH), March 7, 2020; Menominee Tribe of Wisconsin, "Recognition of the Rights of the Menominee River," No. 19-52, https://cswab.org/wp-content/uploads/2020/02/Menominee-Tribe-Resolution-19-52-Recognition of-the rights-of-the-Menominee-River-16-Jan-2020.pdf, 2; Morgan Lowrie, "Quebec River Granted Legal Rights as Part of 'Personhood' Movement," *Gazette* (Montreal, Quebec), March 5, 2021.
53. Maia Wikaira quoted in Smith, "The Klamath River."
54. Stone, "Should Trees Have Standing?," 450–56; Babcock, "A Brook with Legal Rights," 9.
55. City of Chicago, *Chicago River Corridor Development Plan*, 2; "Waterways: Rivers, Lakes, and Streams Act," Illinois General Assembly, 615ILCS 5/7, February 7, 1996, https://www.ilga.gov/legislation/ilcs/ilcs3.asp?Actid=1781&Chapterid=47; Hill, *The Chicago River*, 186; U. S. Army Corps of Engineers, *Partnering with the U.S. Army Corps of Engineers*, 1, 19.
56. Algren, *Chicago*, 73.

BIBLIOGRAPHY

Algren, Nelson. *Chicago: City on the Make*. Chicago: University of Chicago Press, 2011.

———. *The Neon Wilderness*. New York: Seven Stories Press, 1986.

Arnold, Isaac N. *William B. Ogden; And the Early Days in Chicago*. Chicago: Fergus Printing Company, 1881.

Babcock, Hope M. "A Brook with Legal Rights: The Rights of Nature in Court." *Ecology Law Quarterly* 43, no. 1 (2016): 1–51.

Black Hawk. *Black Hawk's Autobiography: Through the Interpretation of Antoine LeClaire*. Edited by J. B. Patterson. Rock Island IL: American Publishing Company, 1912.

Bluestone, Daniel. *Constructing Chicago*. New Haven CT: Yale University Press, 1991.

Cain, Louis P. *Sanitation Strategy for a Lakefront Metropolis: The Case of Chicago*. DeKalb: Northern Illinois Press, 1978.

Chrzastowski, Michael J. "Chicagoland: Geology and the Making of a Metropolis." Champaign: Illinois State Geological Survey, 2005.

City of Chicago. *Chicago River Corridor Development Plan*. Chicago: Department of Planning and Development, 1999.

Coclanis, Peter A. "Business of Chicago." In *Chicago Business and Industry: From Fur Trade to E-Commerce*, edited by Janice L. Reiff, 42–44. Chicago: University of Chicago Press, 2013.

Cronon, William. *Nature's Metropolis: Chicago and the Great West*. New York: W. W. Norton & Company, 1991.

Crutzen, Paul F., and Eugene F. Stoermer. "The Anthropocene." *Global Change Newsletter* 41 (May 2000): 17–18.

Doyle, Martin. *The Source: How Rivers Made America and America Remade Its Rivers*. New York: W. W. Norton & Company, 2018.

Dryzek, John. *The Politics of the Earth: Environmental Discourses*. 4th ed. Oxford: Oxford University Press, 2022.

Ferguson, Arthur B., Jr., and William P. Bryson. "Mineral King: A Case Study in Forest Service Decision Making." *Ecology Law Quarterly* 2, no. 3 (Summer 1972): 493–531.

Flood Control Coordinating Committee. *The Chicago Underflow Plan: Development of a Flood and Pollution Control Plan for the Chicagoland Area*. Chicago: Metropolitan Sanitary District of Greater Chicago, 1972.

Gerstle, Gary. *Liberty and Coercion: The Paradox of American Government from the Founding to the Present*. Princeton NJ: Princeton University Press, 2015.

Gustaitis, Joseph. *Chicago in 50 Objects*. Charleston SC: History Press, 2021.

Happel, Austin, and Dustin Gallagher. "Chicago's Fish Assemblage Over ~30 Years—More Fish and More Native Species." *Urban Ecosystems* 24 (April 2021): 311–25.

Harpster, Jack. *The Railroad Tycoon Who Built Chicago: A Biography of William B. Ogden*. Carbondale: Southern Illinois University Press, 2009.

Herrick, Robert. *The Gospel of Freedom*. New York: Macmillan Company, 1898.

Hill, Libby. *The Chicago River: A Natural and Unnatural History*. Rev. ed. Carbondale: Southern Illinois University Press, 2019.

Jung, Patrick J. *The Black Hawk War of 1832*. Norman: University of Oklahoma Press, 2007.

Katz, Eric. *Nature as Subject: Human Obligation and Natural Community*. Lanham MD: Rowman & Littlefield Publishers, 1997.

Kramm, Matthias. "When a River Becomes a Person." *Journal of Human Development and Capabilities* 21, no. 4 (2020): 307–19.

Latrobe, Charles Joseph. *The Rambler in North America*. Vol. 1. New York: Harper & Brothers, 1835.

Margry, Pierre. *Découvertes et éstablissements des français, dans l'ouest et dans le sud, de l'Amérique Septentrionals (1614–1754): Mémoires et documents originaux recueillis et publiés*. Vol 1. Paris: Imprimerie D. Jouast, 1875. Translated by Shannon Bailey, August 5, 2021.

McCafferty, Michael. "A Fresh Look at the Place Name Chicago." *Journal of the Illinois State Historical Society* 96, no. 2 (Summer 2003): 116–29.

Parsons, John W. *History of Salmon in the Great Lakes*. Technical Paper no. 68 of the Bureau of Sport Fisheries and Wildlife. Washington DC: U.S. Government Printing Office, 1973.

Ralph, Julian. *Our Great West: A Study of the Present Conditions and Future Possibilities of the New Commonwealths and Capitals of the United States*. New York: Harper & Brothers Publishers, 1893.

Rose, William E., and Associates. *The Chicago Portage and Laughton Trading Post Area: "The Waterway West."* River Forest IL: Forest Preserve District of Cook County, 1975.

Rosenzweig, Michael L. *Win-Win Ecology: How the Earth's Species Can Survive in the Midst of Human Enterprise*. New York: Oxford University Press, 2003.

Sandburg, Carl. "Chicago." In *Chicago Poems*, 3–4. New York: Henry Holt and Company, 1916.

Sanitary District of Chicago. *Proceedings of the Board of Trustees of the Sanitary District of Chicago from January 1, 1900 to December 31, 1900*. Chicago: John F. Higgins Printer, 1901.

———. *Report on the Sanitary and Ship Canal*. Chicago: The District, 1906.

Scarborough, William, Faith R. Kares, Iván Arenas, and Amanda E. Lewis. *Adversity and Resiliency for Chicago's First: The State of Racial Justice for American Indian Chicagoans*. Chicago: Institute for Research on Race and Public Policy at University of Illinois at Chicago, 2019.

Sowards, Adam M. *The Environmental Justice: William O. Douglas and American Conservation*. Corvallis: Oregon State University Press, 2009.

Stone, Christopher D. *Should Trees Have Standing? Law, Morality, and the Environment*. 3rd ed. New York: Oxford University Press, 2010.

———. "Should Trees Have Standing? Toward Legal Rights for Natural Objects." *Southern California Law Review* 45 (1972): 450–501.

Telander, Rick. "Salmon for Hizzoner." *Sports Illustrated* 46, no. 14 (March 28, 1977): 60, 63.

Thomas, Jesse B. "Statistics Concerning the City of Chicago: Report of Hon. Jesse B. Thomas as a Member of the Executive Committee Appointed by the Chicago-

River-and-Harbor Convention." In *Chicago River and Harbor Convention*, edited by Robert Fergus, 175–96. Chicago: Fergus Printing Company, 1847.

Thwaites, Reuben Gold. *The Story of the Black Hawk War*. Madison: State Historical Society of Wisconsin, 1892.

U.S. Army Corps of Engineers. *Partnering with the U.S. Army Corps of Engineers: A Guide for Communities, Local Governments, States, Tribes, and Non-Governmental Organizations*. Alexandria VA: U.S. Army Engineer Institute for Water Resources, 2019.

White, Richard. *The Middle Ground: Indians, Empires, and Republics in the Great Lakes Region, 1650–1815*. New York: Cambridge University Press, 2011.

Winslow, Charles S., ed. *Indians of the Chicago Region*. Chicago: C. S. Winslow, 1946.

Wolfert, Ira. "The Seven Wonders of American Engineering." *Reader's Digest* 67 (November 1955): 123–30.

Part 4 DELTA

Rewilding's Perils

Part 4 grapples with the side effects of urban rewilding and illustrates how environmental racism is as much a part of the American experience as the frontier. As someone only three generations removed from slavery, Robert Bartlett, in "A River Lament," reminds us that his native West Virginia served as the frontier for African Americans escaping the antebellum South and that their reliance upon urban farming and fishing might well be a vision for the future.

Taking a deep dive into the multigenerational relationship between family, culture, lifestyle, and a small-town river, Bartlett examines what we've lost when our rivers (or childhood creeks) become severed from our communities. While Bartlett's family history is emblematic of the Black Appalachian experience, it is also the history of America's reckoning with race and the environment. Bartlett shows how the events—racial, economic, cultural, and environmental—of the late twentieth century led to a rupture in our relationship with the natural world. Bartlett's narrative serves two functions: it provides a baseline, so that younger generations can understand what was lost, and it also reminds us that the grief in losing places such as New Creek need not prevent us from saving what we can and moving forward to develop a new and more holistic relationship with our rivers and the places in which we settle.

While Bartlett provides a personal account of the loss that stems from river "improvements," C. Ian Stevenson, in "Restoration's Dark Side," explores how political and economic structures lead to displacement of marginalized communities through gentrification. Stevenson raises an important question—who benefits from urban rewilding? Rewilding doesn't happen in a vacuum, as Stevenson reminds us—urban rivers are hybrid ecosystems informed by social relations and are thus often contested spaces. Stevenson shows how environmental improvements on the Saw Mill River in New York disproportionally benefited the middle and upper classes rather than providing increased access to nature for the existing inhabitants. In Yonkers, the urban renewal process was explicitly linked to the revitalization of the Saw Mill River and a replacement of the Black and Latino residents to make way for a new, wealthier population.

Like Stevenson, Bryan Rasmussen, in "Angling in the Anthropocene," argues that the promise of urban rewilding perpetuates "racial injustice by inviting wealth and privilege onto a river that whites once eagerly abandoned." On the Los Angeles River, Rasmussen sees two rivers coexisting at the same time and place—the white river and the brown river. The first is a river of increased recreational opportunities, epitomized by fly-fishing with all its elite trappings, including the seeming incongruous practice of catch and release. In contrast, the brown river is characterized by bait fishing for sustenance, primarily by people of color. Rasmussen shows us how these two rivers converge when fishing for carp, a racialized fish. We meet Lino, a second-generation Filipino from East LA, who guides free fly-fishing outings, and Fernando, an angler from Michoacán, who seeks solitude on the river.

As these chapters, and this volume, demonstrate, rewilding in the Anthropocene is complicated and messy. While there are no easy answers for restoring biodiversity and natural processes along with healing our relationship to the natural world, there is a world of possibilities—a new frontier, if you will. Although the Anthropocene is literally the age of humans, it can also be the age of resilience and restoration, and through this messy swamp of rewilding, it can signal the end of hubris and the beginning of an age when humans and nature can coexist.

11 A River Lament

New Creek and the People Who Loved It Near to Death

Robert Bartlett

As early as I can remember, my step-grandmother, "Aunt Pearl," kept a fishing pole next to her kitchen door for me. The pole was a stripped, tapered branch of birch with kite string for line and a safety pin for a hook. Thin strips of beef bologna served as bait. Aunt Pearl would wrap the bologna with a slice of Wonder Bread or some oyster crackers in newsprint. As I bolted for the creek, across the street, she would call out, "Baby, if the fish aren't bitin', remember you still have lunch." Jokingly, she'd add, "And don't come back without dinner." Aunt Pearl was Grandpa's third wife. Although we weren't related by blood, she loved my brother and me as if we were her own. She was a gentle, light-skinned Black woman with an infectious laugh, who made clothes and did mending for a living.

I later replaced Aunt Pearl's birch sticks with rods and reels handed down from my father and others I ordered from the Sears and Roebuck mail-order catalog. Aunt Pearl's strips of bologna gave way to live bait suspended beneath plastic red-and-white bobbers, then to rubber worms with little propellers in the front and heavy metal spinners, and later to near weightless hooks adorned with fur and feathers. Little did I know then that my adventures on New Creek (pronounced *crick*), only sixty strides from Aunt Pearl's front porch, would grow into a lifetime passion—an obsession really—for fly-fishing and free-flowing rivers.

Within a short stroll of New Creek in Keyser, West Virginia, four generations of my family lived, died, and are buried. Among them were urban

16. The evolution from bait fishing to fly-fishing thanks to Curt Gowdy and *The American Sportsman* show. Depicted are artifacts of a fishing life: bait-casting reel, live-bait can, fly reel, and an artificial fly. Photo courtesy of Robert Bartlett.

farmers, domestics, a school janitor, a seamstress, a delivery driver, a railroad laborer, an assistant undertaker, and a Civil War veteran. They were hard-working, deeply patriotic, small-town, country folks who knew how to live off the land. They loved to hunt and fish—skills purposely passed on to their children. In Keyser in the 1950s and '60s, Black people were of the river and of the land.

My river story is a collection of memories—memories of joy, loss, regret, and sorrow. It is a lament not just for what once was, but for what still is: creeks choked by earth and rock-filled dams, upstream watersheds struggling to survive, and a people gone from the waters that long sustained them. While this story began with me and a birch stick, it is also a history of my family, the history of a small community of Black folks, a history of place and of our home water, once a wild tributary to the North Branch Potomac River. It is a river tale of relationships and consequences in a place of both wild natural beauty and environmental degradation. My story is that of a people caught in between.

In West Virginia, commerce came with consequences. Coal mining flourished with the arrival of the railroad in 1843, and company towns for the coal mines and paper mills followed. In 1888, five miles upstream from Keyser, a pulp and paper mill was built on the North Branch between the mill towns of Piedmont, West Virginia, and Luke, Maryland. The mill always produced a terrific stink, the result of the noxious chemicals that infected the air. Downwind on the West Virginia side of the mill, the toxic ash denuded the mountainside. Locals who worked at the mill, some for generations, defended the foul smell, claiming, "That's the smell of money." Yet, the U.S. Environmental Protection Agency and others have listed the paper mill as one of the largest emitters of toxins in Maryland.

From the south, New Creek flows into Keyser, where its clean water merges in town with the toxic flow of the much larger North Branch. Thick and brown, the North Branch reeked of sulfides, ammonia, and wastewater discharged from the mill, while toxins stained its banks a rusty orange. The North Branch continues downstream into the Potomac River, a river highway once filled with barges of coal, iron ore, and timber extracted from the Allegheny Mountains upstream. We never fished the North Branch.

In 1830 the Baltimore and Ohio Railroad (B&O) pushed west from Washington DC. Railroad trestles spanned the Potomac along the route, while roads and towns, Keyser among them, rose alongside. With the onset of the Civil War, the mountain people of Virginia sided with the Union and separated from the slaveholding eastern region to become West Virginia. The Potomac River became the border.

Slavery brought my family inland from the coast to the mountains to New Creek. My father's earliest-known relative was purchased from a slave block on the Virginia coast as a young girl and brought inland tied to a wagon. My mother's earliest relative to arrive in the area was a runaway enslaved person named John. Following the Emancipation Proclamation in 1863, my great-great-grandfather on my father's side, Andrew Tross, left Keyser for Pennsylvania to enlist in Company K, 127th United States Colored Infantry. After the war, he was medically discharged and lived out his life in Keyser. He died in 1910 and was buried in the single cemetery in our county reserved for Black folks.

Although West Virginians fought for the North, segregation still ruled for another century. Keyser had a designated Black school and two Black churches, Methodist and Pentecostal. My father was born into the Methodist Church and my mother into the Pentecostal. I don't remember any white families in either church. Mom and Pop grew up only a short distance apart and were schoolmates in the 1930s at the same "Negro-only" schools—Lincoln Elementary in Keyser and Howard High School in neighboring Piedmont. Black people in Keyser bought property and formed small racial enclaves in an otherwise white town with segregationist attitudes. Black folks in Keyser developed ties to one another and the land. Even within the city limits, my family and other Black folks kept large gardens, dairy cows, pigs, chickens, and horses. They were hunters and fishers who often vacillated between abusing the land and waterways to treating them as sacred.

My great-grandparents' two-story house marked the beginning of a row of Black families living side-by-side along New Creek where the water's edge was a stone's throw from their front porches. The two-hundred block of Water Street consisted of untamed front yards that bordered a poorly maintained city sidewalk. Water Street was then, and still is, a one-way single lane of asphalt with a two-foot-high floodwall separating humans and the water. The concrete barrier marked the end of civilization. Crossing the sidewalk, the street, then climbing solo over the floodwall with my birch pole and bologna marked a rite of passage for me.

For my relatives and their Black neighbors, raising kids along the creek and living with its natural dangers and spring floods was part of the natural order of things. In addition to floods, the creek bank was home to beavers, muskrats, frogs, turtles, lizards, snakes, and plenty of poison ivy. The creek's shoreline housed a variety of shore birds, like grebes, plovers, and sandpipers. Resident wild and domestic ducks and geese nested there. It was their home water, too.

Aunt Pearl's place was more a cottage than a house, seemingly too small for raising a family. Weather permitting, visitors sat outside. My favorite place was the screened-in front porch. During summer months, the screens were covered in wild, sweet-smelling honeysuckle shrubs that attracted bees, butterflies, and hummingbirds. There was a constant summertime porch buzz, and the view of the creek from the porch entertained me for hours. I remember her house as drafty inside, especially during a strong wind. Their roof leaked during a heavy rain or snow, and the living room always smelled damp, yet Aunt Pearl's kitchen always smelled good, with its aroma of fresh-brewed coffee and whatever was on the menu, often fried fish. There was always coffee for adult company and a glass jar full of hard butterscotch or peppermint candy for younger guests. No one ever found the candy jar or Aunt Pearl's coffee pot empty. Her kitchen, where she kept my good fishing sticks, was a tight space. Only two steps from the back door, the dinner table took up most of the tiny kitchen. It was metal and could accommodate only two chairs, one for Aunt Pearl and one for my grandfather, Daddy Frank, whose grandfather was Andrew Tross, the Civil War vet.

Daddy Frank was born and raised in the two-story house next door, which still stands. The house had more yard in the back than in the front, and the creek ran close to the front door. It has been bought and sold multiple times over the years but only to Black folks. Daddy Frank's father, Pop Walter, and his second wife, Ms. Alice, died long before I was born. Pop Walter was known for his love of horses. As a hostler, he kept two in the barn behind his house. He used the horse team and a wagon for deliveries around town. He often drove his team up to my family home where he would pick up my older brother and take him along on his delivery runs. Only one picture of Ms. Alice, dressed in fur, survives, claiming a special place in the family photo collection.

Lining South Water Street on the creek side were giant deciduous trees, the source of my good fishin' sticks: white ash, basswood, birch, black walnut, chestnut, black and scarlet oak, hickory, locust, sugar maple, red sycamore, and an occasional weeping willow. The trees stood guard on the other side of the floodwall. Come fall, leaves of bright orange, red, and yellow painted the landscape a rich mosaic of colors that can still take my breath away. Thick, tree-climbing vines spiraled up their trunks holding their hosts in a tight embrace. Tree trunks and limbs provided habitat for the squirrels, birds, and other creatures that lived along the creek, as well as hiding places for little Black boys. The leafy canopy provided cool shade for me and my childhood friends whether climbing, swimming, or fishing.

New Creek ran clear and cool most of the year, maintaining a healthy river smell of streamside flowers and honeysuckle, ferns, rushes, and river moss. It ran free of coal-mine and paper-mill waste. Its shallow runs, pools, and back eddies were home to a diverse fish population. But heavy spring rains and swollen mountain-feeder streams regularly turned New Creek a deep mocha, and its unencumbered volume burst over the floodwall. Those of us who lived on higher ground would be summoned to South Water Street to help push the creek and the mud and all kinds of creepy-crawly things out of my grandparents' living room. The river sediment left behind enriched gardens, and random deposits of driftwood became yard art.

Human inhabitants and those of the untamed creek coexisted back then. My family cleaned up after floods, and once the water subsided, they took what they needed from the creek until it flooded again. Most importantly, they never tried to tame the creek or protect their homes from it. Floods were nothing personal, simply an accepted consequence of dwelling close to the creek. Property was also affordable in New Creek's urban floodplain. Black folks knew how to counteract economic poverty by sharing the abundance that came from the land and the creek. A healthy, unpredictable New Creek contributed to their sense of well-being.

Life on higher ground was different, where folks never had to worry about creek floods when the heavy spring rains came to the valley. However, there was a large earth-and-rock flood dam higher up the hill. Black folks on Radical Hill and Lincoln Street were crop farmers and raised animals for milk, eggs, and meat. They also foraged for wild fruit, berries, and roots that flourished nearby. The backyards of the mostly two-story houses ended against a steep, densely wooded hillside, which provided plenty of good tree-swinging vines and game trails. Much like on South Water Street, multigenerational families lined the last block of Lincoln Street. There was not a single white person on that block.

Next to the last house on the block was the house where my father, Pop, was born in 1921. Pop's parents, Daddy Frank and Gramma Bertha Lee, who was also born in the same house, shared the home with her parents—three generations under one roof.

Daddy Frank and Bertha Lee married when she was eighteen. They were related by blood, too close some said, and both looked more white than black. Bertha Lee was an attractive, light-skinned woman with dimples and dark, shoulder-length hair, while Daddy Frank was short and stocky. At twenty-three Bertha Lee died giving birth to their fourth child. She died in her home on Lincoln Street in the arms of her mother and a Black midwife. Her four children had come in quick succession, which likely contributed to her death. Daddy Frank, now a stunned twenty-four-year-old widower, divided his children among his wife's parents on Lincoln Street and his own parents on South Water Street. My father and his baby sister were raised on Lincoln Street by their mother's parents. Their siblings were raised on South Water Street by Pop Walter, his first wife, Hattie, who also died young, and then his second wife, Alice.

Pop's childhood home was a large, two-story house with a large garden plot, room for dairy cows, and a root cellar for stockpiling food. According to my father, his grandparents, Pop Burton and Fannie, had milk cows, chickens, rabbits, pigeons, fruit trees, berry bushes, a large grape arbor, and a garden. Pop Burton was also a janitor at the college in town. Pop often told stories of Pop Burton and Fannie and their respect for what they grew and raised, as well as their uneasy truce with the natural wilds behind their house.

Pop Burton and Fannie also died before I was born, but their son Bertus, along with his wife and daughter, continued to live in the house on Lincoln Street. As a child visiting my cousin, I played in the house Pop grew up in. I learned to dance in the same parlor he did and enjoyed running free on the same family property. The chicken coop and rabbit hutch of his youth were still there along with the same outhouse, which I also visited. Indoor bathrooms and city water came late to folks on Lincoln Street.

A few years after the end of the Second World War, Pop bought a little piece of land nearby and two years later finished building our home on it. It was my childhood home from age three and still stands. Like my father, I spent most of my time growing up and playing on Lincoln Street and South Water Street. My family, much like the other Black families we knew, grew deep family roots wherever they settled. My generation was more restless.

Pop stood just over six feet tall with a runner's lean build, and his hazel eyes offset his black wavy hair. Like both of his parents, he had a light

17. Pop fishing on New Creek after work, 1954. Notice the collapsed suspension bridge on the opposite bank. Photo courtesy of Robert Bartlett.

enough complexion to pass for white, but he never tried. He was a Black man who loved the country life, rivers, and mountains, and like all the Black men that I knew, he was tough and confident, at home in nature.

A few of Pop's childhood friends were his closest hunting and fishing buddies. By the time my brother and I were old enough to join them on the river or in the mountains, some were local legends. The best outdoorsman of them all, Mr. Ed, gained celebrity status because of his hunting and fishing skills, as well as his generosity. He could "will" fish into existence, catching them in the most unlikely of places. I saw him do it more than once.

Pop and Mr. Ed were at home on free-flowing water. I remember the two of them creekside, stalking fish like predators. Dressed in rubber hip waders with bait tins fastened to their belts and wielding store-bought fishing poles, they waded in and out, sometimes crossing the creek and other times perching on rocks midstream like blue herons as they contemplated their next moves. Time and distance disappeared. Mom never worried about them, unless one or both of us boys were along. She knew that Pop and Mr. Ed ignored time and caution in equal measure.

Mr. Ed had a small farm at the end of Rat Tail Road along the North Branch in Piedmont. The road to his place was nothing like a rat's tail. We had two different kinds of rats, and neither had crooked tails. Rat Tail Road was as steep and crooked as a dog's hind leg. The road was dirt and rocky from top to bottom—the kind of rocks that caused flat tires. Any visit to Mr. Ed's place was an adventure. Rat Tail was not the kind of road you traveled without a specific purpose, like picking sweet corn at Mr. Ed's farm or taking home fresh eggs by the dozen. Mr. Ed and others like him would give away baskets of farm produce or buckets of fish to visiting Black folks. To white folks he would sell the same produce. The North Branch bordered Mr. Ed's property. He never fished it, and as kids we never went near it. It was dirty and smelled bad.

Pop and Mr. Ed's fishing trips to New Creek or other places yielded stringers loaded with multiple species like catfish, freshwater eel, bass, and sometimes a snapping turtle as well as less hardy river species like trout, which surrendered to death much quicker than catfish, eels, and turtles. Pop and I fished the same water, but Mr. Ed was just so much better at it. When Pop and I fished the creek, I had the feeling he forgot my youthful age and size. He was tall, thin, and agile, and I was not. He just assumed that I could keep up. I developed an unfortunate streamside habit of stumbling and falling in, sometimes over my head. Pop loved to tell stories about my clumsy streamside acrobatics. He always teased me about my lack of river grace. Falling into running water is a habit I have yet to grow out of. I miss Pop teasing me about it.

Mr. Ed passed away soon after I left high school, and as Pop aged, his confidence around moving water faded. He became less sure of himself, and Mom grew anxious whenever he fished alone. New Creek had also changed by then. Earth and rock-filled dams now interrupted the water flow, and the fishing was not as good. Pop, then in his eighties, turned to fishing farm ponds, small ones he could easily walk around.

18. Family outing with cousins at the bottom of Rat Tail Road along the North Branch of the Potomac. Mom (*extreme left*), Pop (*upper right*), brother (*left*), cousin Peachy, Mr. Ed (*in hat, Pop's river buddy*), cousin Ranny (*with stick*), the author (*front and center*). Photo courtesy of Robert Bartlett.

During one of my last summer trips back home, he and I spent a day on one of those ponds because mom wanted fish for dinner. The fishing was not the same as on our creek adventures, but it was good to be with him near water. I did not fall in that day, but I did somehow end up with wet feet and pant legs, which delighted him, and we caught fish. A month before he died, Pop happily fished that same pond.

At eighty-eight and after more than sixty-eight years of marriage, Pop suffered and survived a massive heart attack. Still in the hospital, he suffered a second heart attack while in recovery from the first. I lived (and still do) across the country in the Pacific Northwest and made the coast-to-coast trip in time to say goodbye. We had some quality time together for a few days before hospice took over and the "Do Not Resuscitate" orders were posted around the house. Pop's fishing poles, tacklebox, and belt-fitting bait-can would soon be mine.

When Pop asked me to take him for a short drive, I knew where he wanted to go. Less than five minutes from home, we parked at a familiar bend along New Creek under an ageless weeping willow. The willow looked the same as when I was a kid, only smaller to my adult eyes. The tree stands at a bend in the creek where Pop and I fished countless times. We sat in silence, staring at the creek, listening to the songs of red-winged blackbirds, cardinals, and warblers and the constant buzzing of cicadas. We avoided eye contact, neither wanting to see the other cry. Pop's time on this earth was short, and we both knew it. We knew this visit would be the last time for us to be together in that place. After taking a shallow breath, he said, "Take me home." As I turned my eyes from the water, I said silent goodbyes to my father and to that sacred place. Those moments mark some of my last memories of Pop and the place he taught his two boys to love so much.

Pop passed away quietly in his sleep at home, two blocks south of Lincoln Street where he entered this world and a few blocks west of New Creek. He lived long enough to see dams come to the valley. He never liked them or the lakes they created. He taught my brother and me everything he knew about rivers and mountains. We never fished lakes created by dams nor owned a boat. When not fishing New Creek, we fished other tributaries of the Potomac. He always said, "I just don't have the patience to fish still water." One of my favorite pictures is of Pop, my brother, and me about age ten, standing together streamside proudly displaying a stringer of fish. You can tell by the look on Pop's face that he was proud of his sons and the river skills he was passing on.

Mom's people lived on higher ground, in the south end of the valley on Radical Hill, where the highest concentration of Black folks in Keyser lived. Houses were spread out and the streets less cared for. The Hill had more open land than the creek bottom or Lincoln Street with acres alive with wild cottontail rabbits, bobwhite quail, groundhogs, whitetail deer, wild berries, and good-eatin' and good-throwin' apples. There were plenty of wild greens to pick and sassafras roots to dig for tea. There were a few pig farms, too, and everyone seemed to have chickens. On Radical Hill every living thing on the land had a useful purpose.

Whereas my father's people carried the blood of master and enslaved person in their lighter skins, Mom's family on Radical Hill showed no such signs of willing or forced racial indiscretion. My mother's immediate family members were dark. Short like her own mother, Mom stands barely five feet tall, with smooth, dark skin and hair that needed a hot comb to straighten during her youth. Mom was the second of three children born at home but the only one to survive. Betty, the first, was stillborn, and James died of whooping cough shortly before his first birthday.

Mom is the granddaughter of a runaway enslaved person. Mom's father, John, was born across the North Branch in Maryland twenty years after the Civil War. Years later when he met and married Mom's mother, Ethel, they moved south across the North Branch to Radical Hill. I called her Grandmom, while he was Grandpap.

Grandpap was a hard-working, tireless storyteller full of "good-telling" stories and ageless wisdom. He remembered World War I and the Great Depression like they were yesterday. Grandpap lived for nearly a century, always marveling at airplanes and astronauts landing on the moon. He never trusted banks, especially after the great Wall Street Crash of 1929. He preferred storing his money under his bed mattress or in a sailor's trunk that sat at the foot of his bed. Grandpap guarded his money. He didn't like parting with any of it.

Grandpap loved my mother, his only child, deeply and unconditionally, and that love extended to his two grandsons. Whenever we were together, I begged him for a story, and he usually complied. Once, when I brought my wife and our one-year-old daughter to visit, we sat on his back porch. He was holding his great-granddaughter for the first time when he shared a hidden family story, the story of his runaway enslaved father, John, and his brothers, Bethlehem and Moses. Listening to him, with my young family beside me, was an unforgettable experience.

According to Grandpap, his father and uncles were owned by a family in Virginia. Always yearning for freedom, they finally hatched a plan and ran. Enslaved people knew the consequences of running, and the risks they took were painful for my wife and me to hear. John was the youngest and couldn't keep up with his older brothers, so they left him behind at a safe house in Cumberland, Maryland. They never met or saw one another again. My great-grandfather John lived out his life in Cumberland, and his older brothers lived out their lives in Pennsylvania. As I listened to the story for the first time, the distance between myself and slavery shrank. Grandpap vividly described the escape and reality of runaways as told to him by his father. Grandpap's words brought our horrific family history crashing home.

The fact that Grandpap's father never saw Bethlehem and Moses again made their hardship and survival even more real to me. My wife and I cried as we listened. I found out later that I was the first in the family to hear that sacred story. He had never before shared it, not even with his only child.

Grandpap was dark, handsome, with a strong, lean build, and bald on top with snow-white hair along the sides. He worked as a laborer with the B&O Railroad, laying track as a member of a work gang known as gandy dancers, or section hands. He also worked unloading coal cars by hand. Among our family photos is a picture of him as a young, bare-knuckles boxer. I remember his large hands, the size of bone-in hams. He was also an urban farmer with chickens, ducks, and rabbits in his backyard. On his days off from the railroad, he was a popular butcher for his neighbors on the hill. He killed and butchered pigs for white folks, who often paid him in cuts of meat, often the least desirable parts. I remember my Grandmom cooking pig intestines (chitlins). The smell will never leave me. Grandpap used to say, "Son, we eat everything but the squeal." Black folks paid him in live chickens or farm produce. Grandpap preferred this method of payment much better than cash. Grandpap lived ninety-eight years, illiterate until the day he died. He never had a driver's license or owned a car. He smoked a pipe, rubbed snuff, and drank whiskey but only on payday. He was not a fisherman.

Grandmom, his wife, was short, with a medium build, dark-brown skin, and thick, black hair that never turned gray. She used a red-hot metal comb to straighten her hair, the same one she used on mom. Grandmom had a body made for hugs and a lap perfect for cuddling. A hug from her was a full-body experience. Her hands grew chapped and rough in winter from splitting wood and hauling coal for the heat stove in the living room and the cook stove in the kitchen. Grandmom wore a dress with a duster (a full-length apron) over it and a hairnet on her head, except on Sundays when she dressed up for church. On Sundays her hairnets gave way to beautiful felt hats adorned with feathers.

Grandmom was a Pentecostal, a daily Bible reader, and a faithful churchgoer. She was a "Holy Roller." Her all-Black congregation believed that Sunday church was an all-day affair. Grandpap never went. Summer Sundays inside the small church were hot and humid. With its pews full, the entire place smelled of sweaty bodies, the bodies of men dressed in dark suits that absorbed the heat and women dressed in their finest, trying to beat back the heat with handheld paper fans. Most fans featured colorful Bible scenes and printed Scripture. Others had a single portrait printed on both sides of a blond, blue-eyed, white Jesus.

Winter days in church were the opposite with the same space now warmed by fiery preaching and a wood-burning stove, as well as the churchgoers' body heat and enthusiasm. The women of the church were as good at raising God-fearin' children and cooking and baking for potluck Sundays as the men were at hunting and fishing. If you were Black in our valley, your worth was measured by how hard you worked. For folks who worked sunup to sundown six days a week, Sundays were a blessing, set aside for worship, rest, and fishing.

Black women, like Grandmom, raised children, kept the stoves burning, took care of gardens, fed the chickens, collected their eggs, slopped the hogs, and cooked and cleaned. They picked wild greens, mushrooms, ramps (a strong wild onion), and wild berries. They dug sassafras roots, picked dandelions, and turned their flowers into wine. Some worked as teachers at Lincoln Elementary in town or at Howard High in Piedmont. Most worked as nannies, domestics, and midwives. Midwives delivered both Black and white babies, babies that other Black women helped raise.

Mom and Pop started dating in high school and were married in 1942. Soon after Pearl Harbor, Pop volunteered to join the army, while Mom lived with her parents. My older brother, Walter Jr., was born in late 1942 while Pop was away in the army. Like Pop, my brother was born in our grandparents' house. In 1945 at the end of the war, Pop returned to his wife and three-year-old son and the home of his in-laws on Radical Hill. I was born four years later, the first on either side of the family to be born in a hospital. For the first three years of my life, both of my parents, my older brother, and I shared the same small bedroom. I slept next to or in Mom and Pop's bed, and my brother slept next to the wood-and-coal-burning potbelly stove that stood in the corner of the room. Pop found full-time work and began saving to build his own house. In 1952 we moved into the house Pop built on the hill between his childhood home and Mom's.

Sunday afternoons were spent visiting more distant friends and family. Some of Mom's relatives lived less than an hour's drive south and east of our valley. They lived along the railroad tracks and along the banks of the South

19. Mom and Pop while Pop on leave during WWII. Photo courtesy of Robert Bartlett.

Fork of the South Branch of the Potomac River in the town of Moorefield. We often made the drive, stopping to fish on our way there and back.

One of Mom's older relatives (I can't recall her name) lived there. She was a large, dark-skinned woman whose body still bore the scars on her wrists and ankles from cruel former white employers who, when she was young, treated her like an enslaved person. The old woman mentioned this treatment offhandedly. Being kids, we never asked for more details or context. Flowing dangerously close to her backdoor, and occasionally through it, was the South Fork of the South Branch of the Potomac. Of course, she fished it regularly.

Now ninety-eight, Mom recently shared the story of her great-aunt Emzella who also lived along the tracks, a short distance from the South Branch of the Potomac River in Moorefield where the two rivers merge. As Mom remembered: "My Auntie Emzella was quite the fisherman. She would grab her fishing pole and a coffee can full of worms and go fishin' whenever she had the chance. I can still see her all alone heading down the tracks behind her house to the river with her pole and worms, her dress just a flyin'. Married women never wore pants back then, not even when fishin'. She would always come back with a hefty stringer of fish for dinner."

Fishing was not gender restrictive, and our relatives loved their fish freshly caught, split open from stem to stern, bone-in, headless, and deep fried. The entire fish was rolled in heavily seasoned flour or breadcrumbs, then immersed in hot bacon grease in a cast-iron skillet. Hot bubbling grease turned the fish a crispy brown. There was cornbread, collard greens or green beans cooked with pork, potato salad with bacon, plenty of bread and butter, dandelion wine for the adults, and chocolate cake smothered in chocolate icing for dessert. Before dinner, my brother and I would play with our cousins and their friends along the tracks and the two rivers. The adults trusted us to make safe choices. Our Moorefield relatives were even more country than we were! I would fish with the cousins, while my older brother busied himself pursuing the girls.

For generations Black men raised their sons to live off the land. Men owned as many guns as fishing poles. The openings of hunting and fishing seasons were legitimate skip-school days. In the fall we roamed the creek or the mountains near town before sunup and still would manage to arrive at school before the opening bell.

We played outdoors as kids, just as our parents did before us, and like them, we often flirted with danger. Our parents and grandparents believed the heavily forested mountains or New Creek, even during times of high water and cold winds, would never hurt us. Pop taught us to steer clear of only two things along the creek: snakes, some venomous like the eastern copperhead, and poison ivy and poison oak. Thick ground vegetation covered the creek's riparian zone where both snakes and ivy thrived. The mountainous valley and the creek were our playground, as well as our outdoor survival classroom.

With its miles of shaded banks, smooth river bottoms, and easy roadside access, New Creek was a special place for us year-round. The creek's flow, temperature, and bottom were ideal for fishing, swimming, and wading, and my friends and I spent hours skipping flat rocks across it. When I was growing up, New Creek flowed unobstructed, wild and clear, through the valley and through Keyser.

The only obstruction on the creek was the diversion dam at the city's water filtration plant a mile upstream of Aunt Pearl and Daddy Frank's house. Creek water was diverted at the diversion dam through the plant and treated with chemicals before entering the city's water lines. Many locals stopped drinking city water when the creek was dammed, claiming they could taste the chemicals in it. Some folks drilled wells on their property, while others preferred to fill water jugs from a pipe that tapped a spring on the side of a nearby mountain. The spring runs year-round, and in winter the ground around the spring turns icy and treacherous, yet locals continue to gather at the pipe with their favorite plastic water jugs. I am not sure if the water there has ever been tested, but it doesn't taste like chemicals, and it flows clear and cold. For many years my parents filled their plastic jugs at the pipe.

The diversion dam at the filtration plant creates a shallow lake that froze solid during the winter, making an ideal ice-skating rink. We played hockey on it with sticks fashioned from half-inch plywood scraps, the blades and handles wrapped in black electrician's tape. Most of us wore figure skates instead of hockey skates because none of us had ever played a legitimate hockey game, and only one of us could skate backwards. It didn't matter because neither the ability to skate nor skates were necessary for our brand of hockey. Goalies played without skates or any kind of protection. We spent most of our time trying to knock one another over, rather than paying attention to the puck. Our games lasted as long as our sticks held out. Plywood splinters and strips of electric tape quickly littered the ice. Skating and hockey were some of the few racially integrated events in town. Other social gatherings were not.

Our use and abuse of New Creek changed with the seasons. To stay warm on winter nights, we burned car tires along the bank when playing late on the ice. The flames sent clouds of black toxic smoke into the night air, and the fire charred the ground leaving burnt rings of rubber and copper on the creek bank. We thought the foul burning smells and burnt rubber rings were only temporary. Little did we know of the lasting consequences of our actions.

We visited the creek for a variety of reasons and took from it freely, just as Pop and his parents had done before us. Sometimes my friends and I visited the creek with a profit motive—wading in with a homemade seine (wire screen netting strung between two sticks), then stirring up the cobblestone and gravel upstream of it. With the help of the current, the seine captured crawfish, shiners (minnows), and hellgrammites (larval dobsonflies). We would catch enough to sell, while holding some back for our own use to supplement the worms and grubs dug out of our gardens for bait.

The filtration plant dam was a good fishing spot and a good place to cross and fish on the other side. We fished up and down the creek for all manner of fish: yellow perch, blue-gilled sunfish, small- and large-mouth bass, suckers, carp, and the occasional brook trout. It was subsistence fishing—we ate whatever we caught. Mom came along when the weather was good. Whenever she caught sight of one of us releasing a fish, she would yell, "What was wrong with that one?" We fished until we had no bait. Intuitively, we considered a stringer of mixed river species a sign of a healthy stream. During hunting season, we used the dam as a wet bridge, a shortcut to game trails leading up the mountain from the water's edge. The mountains were home to gray squirrels, fox squirrels, wild turkey, rabbits, bobwhite quail, and white-tailed deer, all potential dinner fare.

My people experienced the creek in a communal way. The sacred and profane weren't far apart in our valley and sometimes even coexisted at the same spot. For my mother's Pentecostal church, Boggs Hole functioned as a baptismal font. Baptisms in the creek brought converts into new life. Members of the Church of God in Christ and curious onlookers parked along the creek, while a balding preacher in a black suit, white shirt, and skinny black tie, and church women dressed all in white, waded in waist-deep for the dunking. The soon-to-be born-again sinner looked to the heavens in anticipation, while members of the congregation and onlookers sang:

Take me to the water,
Take me to the water,
Take me to the wa-a-ter—to be baptized.

Countless romantic outings also started or ended along the creek. Some of us were likely conceived there, only adding to its sacredness.

On other occasions, Boggs Hole served as a car wash. Pop and his friends would drive into the creek up to the wheel hubs, mix Tide laundry soap with creek water, and start scrubbing. Besides detergent, leaking oil and gas also compounded the level of abuse. Mom would stay in the car, while my brother and I joined our friends upstream for a swim. Other car washers waited their turn. Priority was given to whoever got to Boggs Hole first, Baptizers or car washers—the many would yield to the few.

Fishermen also abused the creek, leaving lunch leftovers and litter along the bank. Sometimes litter helped mark the best fishing spots, just as ropes and hanging vines marked the best swimming holes. In a thousand years, future inhabitants and artifact hunters will likely find soup and coffee tins that once held earthworms, small bait bottles that once held bright-red salmon eggs, and a broken fishing pole or two out in mid-channel. Trash was left without regard, and poles and reels sometimes were tossed in the creek out of frustration when the fishing line knotted or tangled into a bird nest on the reel spool.

Henry (Skippy) Louis Gates Jr., of Harvard and PBS fame, grew up in Piedmont near Mr. Ed's farm. Skippy and I were childhood friends. In *Colored People: A Memoir*, Gates accurately portrayed the valley. He writes, "I never knew colored people anywhere who were crazier about mountains and water, flowers and trees, fishing and hunting. For as long as anyone could remember, we could outhunt, outshoot, and outswim the White boys in the Valley."[1]

Rural Black life coexisted inside a larger white society. Country life was the natural order of things. When we watched the national news on television, we saw civil rights marches involving hundreds, sometimes thousands, of Black people filling the streets of Birmingham and Atlanta, with white policemen on horseback plowing into them, blowing them over with fire hoses, and other times subjecting them to vicious police dogs. The evening news showed other Black people rioting in the streets of Detroit and Chicago while chanting, "Burn, baby, burn!" Black men and women wore black berets, Afros, and dashikis with raised clinched fists, screaming, "Black Power!" It all seemed other worldly, totally foreign to us. As Gates writes,

> The dawn of the civil rights era could be no more than a spectator sport in Piedmont. It was almost like a war being fought overseas. All things considered, white and colored Piedmont got along pretty well in those years, the fifty and the sixties. At least as long as colored people didn't try to sit down in the Cut-Rate or the Rendezvous

> Bar, or eat pizza at Eddies, or buy property, or move into the white neighborhoods, or dance with, date, or dilate upon white people . . . or have a drink at the white VFW, or join the white American Legion, or get loans at the bank, or just generally get out of line. Other than that, colored and white got along pretty well.[2]

Although true, I did not fully understand his burning sarcasm or the gravity of the situation until I left Keyser.

With few exceptions, we owned our property, kept to our country ways, played by the rules, and could be critical of those who didn't. Who would deliberately burn their own home and preach violence? It would be years before many of us knew of or had an urban Black experience or began to understand, for example, that inner-city Black folks did not own their homes as we did and that many lived in crowded places called ghettos or projects. They didn't fish or hunt.

While we didn't know anyone who lived in the projects, each of us knew someone fighting in Vietnam or someone who had died there. Volunteers and draftees from Piedmont and Keyser filled the military ranks. My brother, cousins, and neighbors volunteered, just as our fathers and uncles had done during World War II. I, too, would join the army less than two years after high school.

Television in the 1960s was not all about the civil rights movement, rioting, and the war. In 1964 *The American Sportsman*, an outdoor hunting and fishing show hosted by Curt Gowdy, debuted. The show was an unexpected game changer for Pop and me. Gowdy and his guests fly-fished all over the world but mostly on big western rivers in Wyoming, Montana, Colorado, and Idaho. They always caught lots of big fish, mostly trout. Gowdy and his guests effortlessly cast beautifully hand-tied flies that fooled the fish. The places he and his guests fished and the rods and flies they used did not look the least bit familiar to us. I'm almost certain we never saw a Black person on the show, fishing or otherwise. Pop was willing to overlook those details. He wanted to fish their way, and there were local fish to fool.

Pop had dabbled with fly-fishing a few years before the show aired. He found a bait-casting rod that could double as a fly rod. I remember it was a telescoping metal one, longer and stiffer than his other rods. Pop's objective was to learn how to spend more time fishing, especially when the natural bait of worms, grubs, and hellgrammites ran low. For him, fly-fishing was a last resort.

I'm not sure where he acquired his first "real" flyrod, but he owned one soon after *The American Sportsman* began. It was nine feet long, made of split bamboo, and was light as a feather. The reel was the spring-loaded, automatic retrieve variety made by PENN Rod and Reel Co. Pop bought a half-dozen flies at the only sporting goods store in town. The flies were big and bushy with wings. They were specifically tied to float and didn't look much like anything in or along New Creek, but if Gowdy could fool trout with them, so could Pop.

Pop was a quick study and a self-taught fly-fisherman and soon taught me everything he knew about it, which wasn't much. He and I had always carried our own live bait and casting rods when we fished together, but we shared the bait. After watching *The American Sportsman*, Pop started carrying a second rod, his flyrod, and two or three dry flies. When the live bait ran out, we would take turns with his fly rod. To me, casting big, bushy, dry flies came easily and proved to be a gentler, quieter way to fish. Catching and releasing fish caught in the lip extended the experience. Pop still liked keeping some. I was in high school by then, a seasoned, confident fisherman and Pop's regular fishing partner. The river wading distance between us had closed, and on rare occasions I found myself leading him, upstream or down. I still managed to fall in though.

Formed nearly five hundred million years ago, the mountains of West Virginia are dense, mostly round on top, thick with trees and ground-clinging vegetation. The mountains seem to roll on and on. Mountain-born creeks and rivers bubble from them. Water is a constant worry, whether you live in the mountains or between them. Where most of the western mountains are high, sharp on top, and have too little water, the Appalachians have too much. Where dams in the West are made of massive concrete and steel that span deep, rocky canyons, dams in our watersheds are small, made of earth and rock, and designed to slow the flow of mountain springs.

West Virginia dams were constructed to protect towns from floods and the Potomac River from upstream polluters, such as paper mills and coal mines. Dams came to the New Creek watershed in quick succession in the early '50s and '60s. As industry, commerce, and agriculture grew in the valley and along mountain streams, so, too, grew the need to protect people from floods. Dams allowed people to settle in the floodplain and closer to New Creek than ever before.

Watershed planning for New Creek began in 1944 and accelerated under the 1948 Water Pollution Control Act. The 1957 "New Creek-White's Run Sub-Watershed's Work Plan" outlined land treatment and structural measures necessary to achieve maximum practical erosion control and to alleviate flood damage in the valley. Dams were designed to assist in floodwater protection for dwellings, roads, fields, crops, and businesses, and to permit more intensive land use by reducing damage to land in the flood regions. The plan also prioritized making uplands more attractive and productive. Some of the dam's basins were designed for municipal drinking water and recreation, as well as to serve as containment barriers for silt and waste from mines. A total of nine earthen dams were constructed in and near Keyser, one on New Creek and the rest scattered on nearby mountain streams.

Pop's generation watched the construction of the dams and lived long enough for their reservoirs to fill. The dams were built with taxpayer money. However, with few exceptions, once completed, the rights to access their basins, the creek, and water near the dams were returned to property owners and developers who promptly denied access to the public. The New Creek dam, Dam Site 14, severed centuries-old bonds between river people and their river. Dam Site 14 is the largest of the nine dams, the farthest upstream of town and the only one directly blocking New Creek. The other eight dams are within the city limits or within walking distance of each other. Dam Site 14 changed the creek's flow and the downstream fishing forever. Black folks stopped spending much time on the creek after that.

The New Creek watershed dams were built with a life expectancy of fifty years, and now in the twenty-first century, live on well past their expiration date. Sediment and pollution continue to collect behind them. A breach of any single dam would cause more damage downstream than the destruction they were originally designed to prevent. New Creek would breach and flood Water Street and then flow on to downtown Keyser. The North Branch would absorb whatever was in New Creek's path.

By the mid-1970s, dams contributed to the betterment of the Potomac River's water quality. However, people along New Creek began to endure the consequences of water-use restrictions and dangerously low, warm flows during summer months. They also noticed resident fish once common to New Creek were becoming harder to find. Out of necessity, river people became lake people. Familiar swimming, car-washing, and baptizing holes had changed completely. Weakened and over-controlled, the river became a shadow of its old self.

My brother, Walter Jr., joined the Air Force in 1962. After retiring from active duty in the early 1980s, he returned to the valley. Aunt Pearl and Daddy Frank were both deceased, and their small cottage on South Water Street, the one that smelled like coffee and river water, was gone, too. Walter Jr. purchased the vacant lot from Pop and his siblings. He parked a double-wide trailer where our grandparents' house once stood. Unlike our grandparents, the next generation of Black folks who lived on South Water Street took floods as a personal threat. Taming the creek became a means to an end. Their need for protection from floods meant funding dams.

By the spring of 1993, three consecutive floods forced my brother and his family to sell the family's riverfront property to the city. Their Black neighbors and a few white families downstream did the same. Today there is a grassy, city-maintained park where our grandparents' house once stood. Little Black boys no longer scale the floodwall or hide in the trees along the creek. The creek became a managed urban fishery, very different from the waters my Pop, grandparents, and great-grandparents once knew and much different from the river I first began to fish with a stick and bologna.

New Creek remains the main source of drinking water for Keyser. Visitors with rod and reel rely on West Virginia's Department of Natural Resources to dump hatchery-raised rainbow trout into the creek. New Creek is now solely a "put and take," stock-and-kill fishery. The seasonal stocking of non-native fish is an attempt to mediate the loss of native species downstream of Dam Site 14. Hatchery-raised rainbow trout, with no river memories, are introduced both above and below the dam with no expectation of long-term survival. Those who pursue them somehow think stocked trout will fight and taste the same when fried as the native wild species. They don't.

Dams and river-stocking programs provided short-sighted answers, while an entire river culture and a way of life disappeared. The racial enclaves are long since gone and with them, perhaps, the need for an intimate relationship with the creek. Gone are the swimmers, the born-again Pentecostals, and the dirty cars with leaky parts. A Walmart now stands near the bend in the creek where I said my last goodbyes to Pop. The Methodist church remains, but Grandmom's Pentecostal church on Radical Hill is gone. I have heard that kids still ice skate at the Filtration Plant, however.

Pop's love for free-flowing rivers grew in me and set me up for a rude awakening later in life, as stream ecology and environmental science wormed their way into my thought processes and replaced more youthful wonderment. With new understanding, I began to see the abuse and long-term consequences of living with the fantasy that our damaged or altered natural resources will always remain limitless and fixable. We share a dangerous assumption that if we break or ruin our natural resources, advancements in science and technology will heroically fix our mistakes. No harm, no foul. However, few fixes come to our natural world without consequences. We have learned, too late, that nature can be broken, broken beyond easy fixes—and communities can also be broken, beyond repair. Fixing what's broken is an intricate and complicated process.

What has been done to New Creek may be irreversible. An improved ecosystem for New Creek, one even better than before dams, is easier to imagine than the return of a vibrant Black community to the valley. Without financial capital, we were once rich in the way we lived. We had clean water and an abundance of social capital, the kind earned when you invest in relationships with one another, a creek, and public land. Black folks took care of Black folks, and the nearby mountains, the land, and the creek took care of us.

Only a few of my cousins remain in Keyser, none on South Water Street or Lincoln. Most of my generation saw little reason to stay. For myself, I plan to return for a visit in early fall when the mountain foliage begins again to look like a freshly painted mosaic. Decades removed from the Allegheny Mountains and the valley, I still long for those breathless, color-filled autumn moments, for the sound of red-winged blackbirds, cardinals, warblers, cicadas, for familiar places and for my home water. I long for the ancestral connection of place, the people of my youth, and a birch stick. When I return, sweet memories will surely flood my soul, as the creek once flooded my grandparents' home on South Water Street. I mourn the loss of living close to that land as we did, close to family and a wild urban creek. I do not mourn the loss of the imposed segregation of the times. I will forever mourn the loss of the threads that bound us together with unselfish love and an untamed creek (pronounced *crick*).

NOTES

1. Gates, *Colored People*, 16
2. Gates, *Colored People*, 27–28.

BIBLIOGRAPHY

Gates, Henry L., Jr. *Colored People: A Memoir*. New York: Vintage Books, 1994.

Keplinger, Brandon J. *An Adaptive Management Plan for New Creek Lake, Mineral County, WV*. Romney: West Virginia Division of Natural Resources, District II Warmwater Fisheries Management, 2011, unpublished manuscript.

Wolfe, William W. *History of Keyser, West Virginia 1737–1913*. Keyprint, 1974.

12 Restoration's Dark Side

Gentrification and Daylighting the Saw Mill River in Yonkers, New York

C. Ian Stevenson

In December 2010, backhoes broke ground at Larkin Plaza in Yonkers, New York, a city of nearly two hundred thousand people immediately north of New York City. Rather than digging the foundation for a new building, this machinery began the multiphase process to "daylight" the Saw Mill River that had been buried in a culvert since the 1920s when the city turned its back on the waterway. The project, spearheaded by a variety of nonprofit organizations and governmental bodies, aimed to return this Hudson River tributary to its former, flowing glory in a twenty-first-century reenvisioning. Advocates cast the project as simultaneously spurring ecological restoration and economic recovery for a portion of the city long suffering from a combination of industrial decline and rampant crime. Public and private entities believed the daylighted river would solve both problems, and so money poured into its "rewilding," including exposing the current, planting its banks with hearty flora, and installing fish passages to restore its aquatic fauna. This novel landscape, designed to pay homage to the river's historic appearance, would endow residents with a new appreciation for the natural world and bring urban renewal. Yet, a question emerged: who benefits from these environmental and fiscal improvements?

The daylighting of the Saw Mill River reveals the dark side of restoration—how environmental revival can directly promote gentrification. On the surface, rewilding a river seems unambiguously good. It reduces pollution, promotes biodiversity, combats climate change, and adds green

20. Phase one of the daylighting of the Saw Mill River in Yonkers (2010–12) exposed the river and created Van der Donck Park as a form of urban rewilding. Clarence Holmes Photography, Alamy Stock Photo.

space that offers psychologically restorative benefits for the human population. Yet, scholars have increasingly demonstrated that twenty-first-century initiatives must simultaneously focus on the trio of environmental justice, sustainability, and climate change.[1] While cleaning up an urban river seems like an environmental justice victory by reducing pollution exposure, it creates another challenge by raising property values that, in turn, spur gentrification, forcing the residents who might benefit from the river restoration to other locations likely to contain ills the original location just cured. The restoration of the Saw Mill River in downtown Yonkers demonstrates how the conversion of former industrial sites into mixed-use properties coincided with the removal of the lower-income and predominantly non-white population. Thus, for all its gains, river restoration does not always present unequivocal benefits. Instead, rewilding represents another example of how rivers exist as hybrid landscapes, where human culture and nature remain indelibly intertwined, and thus, altering one necessarily affects the other.[2]

Originating in wetlands in Chappaqua, the Saw Mill River wends approximately 23.5 miles entirely within Westchester County, New York, terminating in Yonkers, where it empties into the lower Hudson River. The river's ecological rhythms mirror its surrounding built environments. Within the largely affluent exurban and suburban northern part of the county, the tree-edged Saw Mill supports freshwater fish species like eastern blacknose dace, tessellated darter, trout, and redbreast sunfish. In the last few miles near its terminus in urban, downtown Yonkers, the more tidal estuary hosts anadromous species like American eel and shad. In addition to enduring the ebbs and flows of the parent river, the Saw Mill River feeds the Hudson and is one of the last of sixty-five streams to infuse it with fresh water before it empties into New York Harbor. Despite its relatively small scale, the Saw Mill River has enjoyed outsized attention for its environmental health because of its relationship with the Hudson River.

The modern effort to restore the lower Saw Mill River stems from the environmental movement of the 1960s and 1970s, when awareness of riparian health began to garner attention. Larger rivers, people rightly understood in the wake of the Clean Water Act of 1972, could only become as healthy as the tributaries that nourish it. As early as 1977, local advocates proposed interventions to rewild the largely forgotten river. This movement gained little traction, in part because of the surrounding urban decay, until powerful allies in both the public and private sectors realized the potential of ecological efforts, beyond their strictly environmental impact, for a new vision of Yonkers as a twenty-first-century urban oasis. Only when monied interests involved themselves, did the rebirth of the Saw Mill River take center stage.

From Indigenous use to restoration, the Saw Mill River represents a microcosm of the American river story. Prior to European contact, the Wecquaesgeek, a Munsee dialect group of the Wappinger, established fishing weirs at the mouth of the river, where it meets the Hudson, to catch anadromous species such as eel, shad, and alewives. Such use endowed the river with the name Nepperhan, an Anglicization of the Munsee word *appehan*, meaning "a trap, snare, or gin." Nepperhan continued to be used throughout the nineteenth century interchangeably with Saw Mill.[3] At the river's confluence with the Hudson, the Wecquaesgeek harvested oysters and just upstream angled for freshwater species like sunfish and hunted waterfowl.[4] Beavers impounded the river, creating pools lined with sedges and marsh grasses, yielding an ecosystem managed by people and fauna alike.

In the first half of the seventeenth century, the Dutch displaced the Wecquaesgeek and began harvesting the river for power generation instead of calories. In the wake of Henry Hudson's 1609 sighting of the Saw Mill River, the New Netherlands colony granted an estate to Dutch patroon Adriaen van der Donck, whose honorific title "Jonkheer" made him the namesake of the city of Yonkers. Between 1646 and 1650, van der Donck established a modest estate replete with a sawmill that used the Saw Mill River's ninety-foot drop adjacent to the navigable Hudson River.[5] This infrastructure allowed van der Donck himself to dub the river in 1653 the Saeck Kill (Dutch translation of "saw mill"), reflecting its new anthropogenic use, and this name would live alongside and eventually supplant the Munsee-derived Nepperhan, etymologically contributing to Wecquaesgeek displacement. Subsequent Dutch landowners further impounded the river at this site, including the Philipse family, whose 1682 house still stands. In Yonkers the Philipses imposed upon the land a tenant-farmer system that required the near-complete removal of the Wecquaesgeek from the river's surrounds. The Philipses' British allegiance in the Revolutionary War a century later resulted in the loss of their estate and opened the land for development at the advent of the Industrial Revolution, with portions sold as early as 1785.[6]

Power generation, and later, waste disposal, characterized much of the Saw Mill River's nineteenth-century industrial use in Yonkers. By 1813 the lowest mile of the river featured four dams, each successively creating a pond to build water pressure to power mills. By 1835 these four dams powered a grist mill, four saw mills, and a hat manufacturer, and in 1837 a fifth dam was added.[7] By 1867 Yonkers's final and highest millpond—at 89.15 feet above the Hudson—was developed, representing a crescendo in impoundment and industrial fervor.[8] Yet, the Saw Mill River was only ever a minor waterway, and while it did power industry, by the 1870s many of the factories it spurred along its banks converted to steam power. The river's primary function thus became the removal of industrial pollutants and municipal sewage.[9]

21. D. R. Fecit's 1784 illustration of Philipse Manor Hall, with the Saw Mill River in Yonkers as it empties into the Hudson River in the foreground. Courtesy of Historic Hudson Valley, Pocantico Hills, New York (PM.65.866).

In the wake of these new uses and denser urban housing, concerns arose about pollution and public health. As early as 1858, civic leader R. P. Getty reported to the Yonkers Village Board of Trustees that "these ponds are now made the spectacle of all manner of filth and rubbish by those having premises on [their] margins, and . . . giving forth exhalations offensive to the senses, and which must in the nature of things severely impair the healthfulness of the locality."[10] Ameliorative proposals to reduce the river to a narrow channel never came to fruition, and for decades the problem worsened. By 1885 the *New York Times* reported "a large number of business houses are built over the [Nepperhan] creek on spiles, and the odors arising from it are very nauseating. . . . The stream receives the dyes and waste from several large factories and acts as a sort of open sewer for the filth from hundreds of dwellings and tenements."[11] In response to rife pollution, one local poet in 1891 dubbed it "a snake-like yellow scrawl of scum."[12] In less than a century, the Saw Mill River had transitioned from a driving force of development to a public embarrassment.

This new public health sentiment fostered a proto-rewilding remedy. In 1892 the State Board of Health of New York examined the condition of the river, with Yonkers city officials reporting that sewage was permitted to flow into the slow-moving stream, yet the water above the factories was pure. In addition to raw sewage, the sanitary inspector described rotten glue and waste wool coming into the river from factories. Others reported "dye stuffs from various mills, besides filth and dirt of a varied character . . . occasionally you would find dead animals in the pools above the dam."[13] To solve the problem, city officials recommended the removal of the dams to allow the water to flow more freely and thereby flush out the hazardous waste. The Board of Health agreed, and at the conclusion of a legal process, dynamited the dams forming the fourth, fifth, and sixth mill ponds in 1893.[14] Reports the following year reveal the delight of Yonkers residents at the revitalized river: "Since the Nepperhan River has been allowed free passage . . . it has become a clear running brook. Yesterday morning Assessor Edward Underhill discovered a fine bass in it at Chicken Island, in the heart of the city."[15] The city seemed to embrace, if briefly, the benefits of deindustrializing an urban waterway.

While removing dams somewhat helped the river's health, continued abuse through building along its banks and waste disposal led Yonkers to relegate it to unseen culverts to mitigate increasing pollution, forestall flooding, and essentially forget about the river's existence in the early twentieth century. As the city's population ballooned to nearly one hundred thousand by 1915, the Army Corps of Engineers developed a plan to bury the last two thousand feet—roughly six-city-blocks' worth—of the river from Chicken Island to the Hudson. The project began in 1917, subjugating the river into a flume, incorporated into which was the arched bridge over the mill pond at Warburton Avenue as well as other river crossings and overpasses that already acted to visually remove it.[16] This intervention was justified by independent surveys, such as one in 1919 that reported from a hospital "effluent from the filters is discharged in a putrescible state directly into the Saw Mill River, causing a marked discoloration of the stream. . . . A great many privies and cesspools are located on the edge of the Saw Mill River and its tributaries and there is also drainage from poultry yards, barnyards and house drains."[17] In 1922 the work was complete, and the Saw Mill in downtown Yonkers was a river in name only, removed from sight

22. Flume construction to bury the Saw Mill River in Yonkers under Larkin Plaza, 1917. Courtesy of Yonkers Historical Society.

and described that year as "a small and obscure waterway of no particular use and more of a nuisance than anything else," reflecting the sentiment that no one would miss it.[18]

The control and effective removal of the river from the city's landscape and consciousness of its citizens resulted in more environmental deterioration for most of the twentieth century that coincided with the economic growth of industrial Yonkers and subsequent economic divestment downtown. Some residents did lament its burial, with one newspaper in 1953 even calling it "imprisonment" and gloatingly announcing that "not without occasional rebellions does the river stay underground and often it breaks into the basements of stores built above its underground course."[19] A year later, one of the city's largest employers, the Alexander Smith Carpet Mills, closed and moved to Mississippi, inaugurating downtown Yonkers's slow economic decline.[20] As jobs left, so did the predominantly white workers, who fled to more suburban parts of the city and to neighboring communities in Westchester County and the Hudson Valley. This population exodus and lowering of residential property values precipitated an influx of other residents of lower economic means, including disadvantaged African American and Latino populations migrating northward from the southern United States and neighboring New York City. From 1950 to 2010, Yonkers's white population shifted from 96.7 percent to 41.4 percent, while African American and Latino residents increased from 3.2 percent to 16 percent and from "unregistered" to 34.7 percent, respectively.[21] As a result, white Yonkers inhabitants further disparaged the neighborhood, seeing it unworthy for investment.

By the 1970s, new sensibilities stemming from the nascent environmental movement brought renewed attention to the river's health, even if residents did not show concern for the economic situation of its nearby residents. Upstream from Yonkers at places like Ardsley, a series of chemical leaks in the early 1970s led to a five-part newspaper exposé.[22] As a result of this growing environmental awareness, in 1977 Frederick Danback, a trustee of the Hudson River Fisherman's Association (HRFA), first proposed rewilding the Saw Mill River based purely on ecological justifications. In an editorial titled, "Flume Woes," Danback asked readers to envision a return of the river to its "natural state" and suggested renovating the flume, a position not unlike that of other nonprofits focused on restoring fisheries, such as Trout Unlimited's support of dam removal.[23] Coupled with the public plea, Danback and the Yonkers chapter of the HRFA took a direct-action approach, going into the flume and plugging a discharge pipe connected to the Otis Elevator Company plant that was releasing oil, paint, and other detritus directly into the river. When a waste engineer at Otis discovered the backflow, the company found other means of disposal.[24] The HRFA was known for being loud and demanding, yet responsible, and an ensuing public debate planted seeds of change.

Just six months after Danback's opinion piece, Yonkers caught national attention for its citizens' resistance to racial integration centered on the population living near the Saw Mill River downtown. In 1977 the Yonkers Public Schools superintendent proposed closing several underperforming schools with predominantly African American and Latino students and implementing crosstown busing to alleviate declining enrollment but also desegregate the schools. The city's white families resisted the proposal, objecting to busing African American and Latino students to their neighborhood elementary and middle schools, despite federal mandates that prohibited segregation.[25] While the migrations of Latino and African American populations had created a vibrant community, including musical artists like DMX and Mary J. Blige (both moved there in 1975 as very young children), many white, working-class residents of Yonkers—especially in the more suburban neighborhoods defined by single-family homes versus the apartment buildings of downtown—came to view downtown Yonkers as seedy and a place to avoid. Yet, modest efforts at investment did sprout, such as the Yonkers Industrial Development Agency's 1978 purchase of two of the former Alexander Smith Carpet buildings, transforming them into rental units aimed at small businesses for manufacturing or industrial uses.[26] The city tried to revitalize the neighborhood through investment in remaining industry, such as Otis Elevator, which ultimately rejected these efforts in choosing to move its facilities out of state in 1983.[27]

By the 1980s, Yonkers achieved national infamy for its staunch resistance to desegregation, envisioned primarily to uplift the disadvantaged citizens living in the dense urban neighborhoods adjacent to the Saw Mill River downtown. Stemming from investigations into the legality of the busing resistance was a 1985 conviction of Yonkers deliberately engaging in housing segregation.[28] In 1986 federal judge Leonard Sand appointed a federal monitor to oversee the integration of the school district through court-ordered measures and ordered Yonkers to remedy the housing portion of its violations by adopting a plan for low- and moderate-income housing in primarily white neighborhoods. This new proposal meant that the concentration of African American and Latino residents would shift from downtown to a more dispersed model, which frightened white residents even more than the busing proposal. Yonkers city government and its majority-white citizenry fiercely resisted the federal mandate, which received even more national attention than the busing issue and later became dramatized in HBO's 2015 miniseries *Show Me a Hero*.[29] These racially-charged attitudes toward downtown Yonkers manifested themselves in even the placename of the Saw Mill River's vicinity, where "Getty Square," named for the nineteenth-century civic leader who once pleaded to clean up the river, became derisively known as "Ghetto Square."[30]

At the same time that Yonkers mayor Nick Wasicsko was fending off angry white citizens about the housing mandate in the late 1980s, more advocacy for the river itself emerged. In 1988 Bob Walters and Andy

Hudak canoed some twenty-three miles of the river from Pleasantville to the Hudson. Walters was then head of Ferry Sloops Inc., a volunteer environmental organization of 350 members with ties to Clearwater Sloops, Pete Seeger's environmental nonprofit aimed at improving the Hudson River. Ferry Sloops subsequently agreed to "adopt" the Saw Mill and conducted a fish survey to raise public awareness. Walters cited how important tributaries were to the health of the Hudson, which had been garnering attention for its own environmental woes. The organization sought to penalize polluters, including the city of Yonkers, whose Department of Public Works had recently discharged oil directly into the river.[31] This campaign brought more allies—including Scenic Hudson, supported by research from local universities with genuine interest in restoring the river's ecological health—in part because of its impact on the larger Hudson. In 1992 the Army Corps of Engineers gave its approval for daylighting as a valid restoration project, and by the late 1990s, the Corps held public hearings for suggestions of where the river could be restored, with momentum now shifting in favor of uncovering portions of the river in Yonkers.[32] Further allies emerged from Groundwork Hudson Valley (founded 2000) and its subsidiary Saw Mill River Coalition (founded 2001), bringing increased pressure to rewild the Saw Mill River. By the turn of the twenty-first century, then, new ecological sensibilities seemed poised to enact change for this urban waterway.

During previous decades, Yonkers officials had grappled with ways to revitalize the city's downtown, looking for opportunities to marry government funding with private investment. Modest successes, like the Yonkers Industrial Development Agency in 1978, were overshadowed by failures, such as the Otis Elevator experiment of the 1970s. In the twenty-first century, Yonkers would find the formula it sought through the gentrification successes of New York City as both a roadmap and a cascading geographic opportunity. Through environmental improvements touted for their rightful ecological benefits, the city could raise property values via civic investment and zoning-law changes to attract a population priced out of Manhattan, Brooklyn, and other parts of New York City, but whose purchasing power dwarfed that of the current downtown Yonkers population. That is, Yonkers could turn the impacts of New York City's hyper-gentrification into its own efforts, knowing that environmental improvements have been shown to accelerate the process and disproportionately benefit the middle, upper-middle, and aspiring upper classes.

In the first decade of the twenty-first century, the initiative functioned in earnest to transform downtown Yonkers, with the river restoration at the heart of the new vision. While the rewilding was still in planning phases, prescient developers flooded the area, buying up vacant or rundown properties with the plan to build new apartment complexes or renovate historic buildings. Examples of these properties include a three-story, Renaissance Revival trolley barn, converted to lofts and a fitness studio; the Victorian-era pier that came to house a water-taxi service to Manhattan; and a second-story, glass-enclosed, 260-seat restaurant overlooking the Palisades and Hudson River. Coupled with this private investment, major public entities committed huge sums of money. The city of Yonkers undertook a $35 million rebuild of the nearby public library (completed 2002) in a former Otis Elevator building along the first stretch slated for daylighting. Across the street and astride the river itself, the region's Metropolitan Transportation Authority revamped the Beaux-Arts-style Metro-North commuter rail station (completed 2004) at a cost of $43 million.[33] From downtown Yonkers, suddenly, white-collar workers could be at their Manhattan offices in just thirty minutes either by train or ferry, creating a bedroom community. Yet, most of the surrounding area remained economically depressed and in disrepair.

On the heels of these investments to downtown Yonkers's built environment, commitments to rewilding the river materialized, creating a feedback loop of private and public investment. In 2005 New York State's governor George Pataki authorized a critical $25,000 grant for a feasibility study on daylighting the Saw Mill, bringing state money into an otherwise local initiative under the reasoning that it could dovetail with these other improvement projects already underway.[34] In turn, the city of Yonkers invited Groundwork Hudson Valley and the Saw Mill River Coalition to work with the local community from 2005 to 2010 through a Community Benefits Agreement to create concepts for how the new space could be designed for an array of uses including walking paths, shops, and other recreational areas.[35] Furthermore, Groundwork Hudson Valley developed a plan that would integrate ecological habitat along the river—making it a living stream—so that it would not simply be exposed urban water.[36] A new consortium organization named Daylight Yonkers, supported by the Friends of the Saw Mill River, drew widespread appeal from neighbors and businesses to show the state of New York that the local community was behind the initiative. Further funding came from the federal level, including an $889,000 grant awarded in 2008 to Groundwork Hudson Valley by the U.S. Environmental Protection Agency for restoration of the river. The following year, the Saw Mill River Coalition conducted a "BioBlitz" to catalog plants, animals, fungi, and bacteria in the river and its watershed, which yielded more than 650 entries, showing just how important the waterway is to the natural environment.[37] With political support and requisite funding that drew from the investments in downtown Yonkers's economic future, the rewilding of the Saw Mill appeared a reality. The "Great Recession" of 2008–9 slowed efforts both to rewild the river and revamp the downtown, but in 2010 the Yonkers City Council granted $17 million for the daylighting initiative, and the New York Department of Environmental Conservation committed an additional $2 million. The alliance of environmental advocates, private investors, and government funding now existed for a symbiotic goal: clean up the river and clean up the neighborhood.

Phase one of the daylighting, segmented into three components from December 2010 to September 2012, dramatically transformed downtown Yonkers to much fanfare.[38] A former parking lot, Larkin Plaza, became an open-air riverbed as part of a 13,775-square-foot aquatic habitat renamed Van der Donck Park. The new landscape thus honored the city's Dutch namesake, ironically the person who first impounded the river at that site, rather than evoke the area's first settlers, whose experience with the free-flowing river was more analogous to the recently daylighted waterway. The daylight design never attempted to return the river to its precolonial appearance but to restore ecosystem services and function within an urban environment. Thus, engineers diverted the water from the 1920s' concrete flume with a new channel, whose bank included a gentle grade lined with a buffer of native grasses, perennial flowers, and hydrophytic trees. This new eight-hundred-foot stretch of river consists of four ponds and water troughs and empties into the Hudson River in an eddy pool, all of which create habitat for larger fish, such as the American eel and carp. A fish ladder (designed primarily for American eel, white perch, and herring) facilitates upstream passage for smaller fish between the pools, while riprap and a staircase waterfall oxygenate the water as well as provide aesthetic improvement. Meanwhile, a thirty-by-thirty-foot netting chamber traps trash and debris from reaching the river's confluence with the Hudson. Coupled with this, the original flume now provides floodwater overflow, and the surrounding park includes permeable pavers and garden beds that serve as an emergency floodplain. Other cultural amenities, such as an outdoor classroom, benches, mosaics, and a footbridge, have enhanced the overall landscape.[39]

While the Saw Mill River's first daylighting phase undoubtedly enhanced the waterway's ecology and brought aesthetic value, it did not create Yonkers's gentrification as much as it jump-started and then sped it up. Efforts at economic development had been attempted for decades and preceded the authorization of the daylighting, ostensibly placing the rewilding as a stand-alone initiative driven by environmental zeal. Yet, from the beginning of its efforts to rewild, city officials understood its potential with regard to economic impact. By 2012, then, Ned Sullivan, president of Scenic Hudson, could describe the Saw Mill River as "no longer a resource people want to hide," simultaneously characterizing it "not only [as] a catalyst for revitalization of the downtown, but now it will become the centerpiece of the city."[40] Such an attitude found international promotion through the documentary *Lost Rivers* (2012). On the surface, the symbiotic plan appeared a win-win: economic improvement and ecological restoration. But as scholars like Melissa Checker have exposed, the "sustainability myth" is precluded on environmental gentrification.[41] That is, the benefits, as real as they are, were never intended for the local population as much as for a new influx to replace them.

The waterway's beautification was valuable in its own right and would benefit whoever lived in its vicinity, yet the plan was co-opted to promote development geared toward a gentrified displacement of the local population. In 2011, for example, Yonkers changed zoning laws to encourage private development of high-density apartment and office buildings in a formerly industrial area. As part of the sanctioned initiative to alter the area's inhabitants, in 2013 Yonkers closed its nearby city jail and then granted substantial tax breaks to art collector Daniel Wolf and his wife, famed architect Maya Lin, so they could purchase it with the proviso that they would develop it into a private arts space that would hold public events. In effect, the city removed what it saw as an unsavory element and replaced it with one of high economic and cultural value, with Mayor Mike Spano even commenting, "The waterfront should be a place to shop, live, or walk. . . . It's not the right place for a jail."[42] Private groups leveraged the change and precipitated transformations of the urban fabric, such as YoHo, an arts collective that traced its roots to the 1978 purchase of the former Alexander Smith Carpet Mills property, which in 2011 created twenty-five new artist spaces to supplement the fifty studios it had already developed in 2005. In 2014 YoHo helped create Yonkers's first official arts district, having influenced the city to further adjust zoning to allow for retail shops to operate at street level, with the intention of servicing the arts community and those attending the community's regularly scheduled events.[43] Unsurprisingly, the audience for this was largely those living outside the neighborhood.

In tandem with this public assistance for private development, the city itself worked to gentrify the area, implementing a so-called Sixth Borough strategy to lure outpriced New York City residents to their northerly neighbor. In 2012, for example, Mayor Spano promoted the $109 million River Park Center, a 224,000-square-foot mixed-use development, to rise alongside the third phase of the daylighted river, showcasing how the river rewilding itself was now in lockstep with new development. Yonkers sponsored in 2014 the creation of a marketing campaign and economic development office, named Generation Yonkers, aimed specifically at young professionals and couples looking for easy access to New York City but for lower cost than the city itself. Advertisements featured mostly white millennials, rather than local residents, as the target audience, with its campaign goal in 2021 as the desire "to rebrand the Yonkers riverfront which has emerged as a vibrant and diverse community that is attracting new residents, restaurants and shops," clear evidence that the city encouraged gentrification rather than uplift of the current population.[44]

While the first phase of daylighting resulted from decades of environmental advocacy and came after major infrastructural investments in downtown Yonkers, subsequent phases demonstrate how the city used the Saw Mill River's rewilding to enhance its own urban transformation at the expense of existing residents. Phase two, completed in 2016 at a cost of $8.3 million, involved rehabilitation of the one hundred feet of river immediately upstream from Van der Donck Park. While adjacent to the initial rewilding, this stretch was markedly different and featured higher water velocity and actively used buildings astride it. Therefore, planners did not attempt to build a natural bed, but instead constructed a twenty-thousand-square-foot public courtyard from which pedestrians can view the flowing water. With three entrances and surrounded on three sides by buildings, the new Mill Street Courtyard features seating, gardens, ramps over the exposed river, and artwork. Invoking the city's colonial heritage, designers envisioned the courtyard in the *woonerf* (Dutch for "living street") style open to pedestrians and vehicular traffic. Extending this theme, Yonkers commissioned Dutch artist Eelco van den Berg to create a three-story mural highlighting the river's aquatic and aviary fauna. To complement this, the city hired Haifa Bint-Kadi, an Arab American artist living in Yonkers, who designed the mosaic, *The Eel's Journey*, reminiscent of the ecological restoration of the project for fish passage, as well as metal sculptures honoring the area's nineteenth- and twentieth-century industrial workers.[45] This latter decision suggested the city's awareness of needing to represent diversity and acknowledge class issues alongside celebrating ecological restoration. However, months before completion, Empire State Development awarded a $500,000 capital grant for renovations of parts of the former Alexander Smith Carpet Mills for use on yet more galleries, artist studios, and workspaces akin to those pioneered by YoHo.[46] And in 2016 Yonkers offered $7 million in tax breaks to private developers to demolish nineteenth-century buildings adjacent to phase one to begin work on a new $197 million, market-rate, 442-unit housing complex composed of twenty-five-story and seventeen-story towers plus a 539-space parking garage.[47] Yet, these previously abandoned buildings had not been considered for rehabilitation as low-income housing or homeless shelters to supplement the Saw Mill River's daylighting.

Phase three, completed in 2018, only accelerated gentrification at the same time it contributed to the river's rewilding, reflecting the same pattern. Costing $18 million, this phase was more akin to the initial daylighting and aimed to expose more of the river and recreate its natural state.[48] Located where the Saw Mill River makes a sharp bend from a predominantly north-south orientation to the primarily east-west one where it joins the Hudson, this area creates a three-sided peninsula colloquially named Chicken Island based on its nineteenth-century agricultural use. Phase three consisted of channel restoration, such as reinforcing its edges with stone armor, and creation of a 1.25-acre park designed for flood mitigation and embedded with garden beds, permeable pavers, and regraded land. The removal of adjacent storefronts exposed more of the water, and a centerpiece of the park was installation of an undershot water wheel that both paid homage to the river's colonial appearance (see fig. 21) and generates the electricity to power the park's lighting, effectively offering another environmental benefit.[49] Just weeks before completion of the third daylighting phase, Mayor Spano announced that Yonkers would sell a six-acre parking lot on Chicken Island to AMS Acquisitions of New York City. In anticipation of the planned fourth phase immediately adjacent to the parking lot, AMS Acquisitions would develop it into a mixed-use residential center that would include a luxury hotel.[50] This plan replaced an earlier one where the city hoped to construct a minor league baseball park at the site.[51]

As these initiatives demonstrate, the Saw Mill River's daylighting ushered in a phase of intense gentrification for downtown Yonkers, rendering the rewilding benefits not for underrepresented groups. Instead, the evidence points to the top-down approaches by city government, coupled with private investors and environmental advocates, that harken back to earlier Euro-American-centric visions of wilderness intended for white Americans. Other constituents who might have been included in the process, including local African American or Latino populations, or the Lenape Nation—who represent the Native American groups who originally inhabited the area—have been excluded. In 2020 investigative reporter Mansuda Arora probed this intentional oversight in an exposé aimed at asking for whom the river's restoration was intended. He also interviewed local residents, one of whom noted that "they keep saying this is our park. It's not—it's Mayor Mike Spano's park." Local artists feel overlooked in favor of the new avant-garde venues catering to outsiders. In a migration reversal, the largely African American and Latino population that moved to Yonkers following the mid-twentieth-century outmigration of white working-class residents are themselves now being displaced to marginally less expensive communities in nearby Mount Vernon and New Rochelle to make way for largely middle- and upper-class white migrants priced out of New York

City. Homelessness is on the rise in the area, threatening the gentrifying public image that daylighting and property investment intended. With $48 million already spent on the first three phases and a fourth phase at Chicken Island in its planning stages, questions remain about how much more gentrification will spread as the river continues to rewild.[52]

The daylighting of the Saw Mill River reveals how well-meaning efforts to rewild urban rivers can quickly morph into the opportunity to accelerate gentrification disguised as general economic renewal. At the same time that environmental advocates decried the Saw Mill River's putrid state, investment into that area of Yonkers increased by outside developers buoyed by city and state funding. This initially unplanned but ultimately lockstep reflexive partnership meant the daylighting project could contribute to an inner-city oasis upon which to anchor an urban professional settlement to supplant the extant lower-income population. The restoration publicly left ambivalent its specific beneficiaries, but rising costs of living could only attract an upwardly mobile populace. It turns out the economic revival attendant with the ecological restoration in Yonkers is one of displacement rather than providing an interface with nature for the neighborhood's existing citizenry.

NOTES

1. See, for example, Barnett, "Introduction," 3–14.
2. On hybridity with regard to rivers, see Stevenson, "Introducing Environmental History."
3. Similarly, the Wecquaesgeek settlement at this section of the river, Neperhameck, is a combination of *appeh* (meaning to trap) and *amack* or *amuck* (meaning fishing place). Thus, the underlying meaning of the human settlement was "trap-fishing place." Hall, *Philipse Manor Hall*, 27; Shonnard and Spooner, *History of Westchester County*, 47.
4. Van der Donck, *A Description of New Netherland*, 98.
5. Cornell, *Some Reminiscences of the Old Philipse Manor House*, 17.
6. Scharf, *History of Westchester County*, 16.
7. Shonnard and Spooner, *History of Westchester County*, 561.
8. Owen, "Old Mills along the Saw Mill River," 4.
9. Allison, *The History of Yonkers*, 228.
10. Madden, *A Documentary History of Yonkers*, 168.
11. "Breeding Malarial Fevers," *New York Times*, August 17, 1885.

12. "Rising River," *New York Times*, January 30, 2005.
13. Moffit, "Exhibit A," 88.
14. "Great Joy in Yonkers: Removal of the Condemned Dams on the Nepperhan," *New York Times*, April 7, 1893.
15. "Yonkers," *New York Times*, July 25, 1894.
16. Barbara Whitaker, "Yonkers Pursuing Plan to Take the Lid Off a River," *New York Times*, July 1, 2007.
17. State of New York, *Fortieth Annual Report of the State Board of Health*, 363–64.
18. Brown, *Old Yonkers*, 189.
19. "The Nepperhan Flume," *Daily Argus* (Mount Vernon NY), August 24, 1953.
20. This displaced 2,400 workers, itself a decline from the 7,000 employed during World War II. Sales Management, "An Industry Moves South," *Reader's Digest*, March 1956, 27–30.
21. Gibson and Jung, *Historical Census Statistics*; "Marymount College Students Chronicle Migration of Southern Blacks to Westchester County," *Fordham News* (New York City), April 12, 2002, https://news.fordham.edu/university-news/marymount-college-students-chronicle-migration-of-southern-blacks-to-westchester-county/.
22. Wayne Nicholas and Gary Kriss, "Oil Spill Kills Saw Mill Fish," *Reporter Dispatch* (White Plains NY), October 20, 1971; "Oily Film on Saw Mill and New Trout Die," *Reporter Dispatch* (White Plains NY), June 15, 1971; John A. Costa, "A River Dies in Westchester," *Westchester Rockland Newspapers* (NY), January 29, 30, 31, February 1, 2, 1973.
23. Frederick Danbeck, "Flume Woes," *Herald Statesman* (Yonkers NY), February 4, 1977.
24. Bob Walters (former director of Ferry Sloops Inc.), in discussion with the author, August 8, 2011.
25. Lena Williams, "A Busing Proposal Divides Yonkers," *New York Times*, March 19, 1978. See also United States of America and Yonkers Branch-National Association for the Advancement of Colored People v. City of Yonkers, Yonkers Board of Education, and Yonkers Community Development Agency, 856 F.2d 444 (2nd Cir. 1988).
26. Penny Singer, "In Big Factory, Small Businesses Prosper," *New York Times*, August 26, 1990.
27. James Feron, "Otis Elevator to Leave Birthplace," *New York Times*, December 1, 1982. At the time, Otis employed 375 workers, down from its height of 1,300.
28. Lena Williams, "Judge Finds Yonkers Has Segregation Policy," *New York Times*, November 21, 1985.
29. This miniseries was based on a 1999 book of the same title by former *New York Times* writer Lisa Belkin.

30. "Robert P. Getty Dead," *New York Times*, March 29, 1902. On the "Ghetto Square" moniker, see, for example, "What Is Getty Square Like These Days," City Data, August 23, 2017, http://www.city-data.com/forum/westchester-county/2811901-what-getty-square-like-these-days.html.
31. Tessa Melvin, "Canoe Trip Leads to a River's Adoption," *New York Times*, January 15, 1989.
32. Sloan, *Daylighting the Saw Mill River*, 50.
33. Elsa Brenner, "It Had the Setting; Now It Has the Housing," *New York Times*, March 8, 2008; Elsa Brenner, "Restored River a Boon to Yonkers," *New York Times*, August 12, 2012.
34. Kirk Semple, "Yonkers Plans to Uncover River Running through It," *New York Times*, January 6, 2005.
35. Sloan, *Daylighting the Saw Mill River*, 51.
36. Ann-Marie Mitroff (former director of River Programs, Groundwork Hudson Valley/Saw Mill River Coalition), in discussion with the author, July 27, 2011.
37. John Senn, "EPA Grant to Spur Restoration Efforts in New York's Saw Mill River Watershed," Environmental Protection Agency, December 2, 2008, https://www.epa.gov/archive/epapages/newsroom_archive/newsreleases/769f4096353be2f785257513006ae1b8.html; "Saw Mill River BioBlitz," *On the Wing*, 2009, https://www.sawmillriveraudubon.org/wp-content/uploads/simple-file-list/newsletters/2009-05-Audubon_Newsletter_Winter-2009.pdf.
38. John Jordan, "The River Runs through It, Again," *Construction News*, December 1, 2010.
39. Colin Gustafson, "Saw Mill River Celebrates Some 'Daylighting' in Yonkers," *Journal News* (Westchester County NY), December 7, 2011; Sloan, *Daylighting the Saw Mill River*, 51–52.
40. Sullivan, cited in Brenner, "Restored River a Boon."
41. See Checker, *Sustainability Myth*.
42. Cherry Wymer, "Yonkers Seeking Developer to Transform Run-Down Old Jail," *Newsday* (Long Island NY), May 8, 2013.
43. John Golden, "Owners Seek Arts District at Historic Carpet Mills," *Westchester County Business Journal*, August 21, 2014.
44. Elizabeth Bracken-Thompson and Dean Bender, "New Marketing Campaign Celebrates Yonkers' Remarkable Riverfront Renaissance," Generation Yonkers, October 7, 2021, https://generationyonkers.com/press-release/new-marketing-campaign-celebrates-yonkers-remarkable-riverfront-renaissance/#:~:text=The%20goal%20of%20the%20campaign,tagline%20Go%20Anywhere%20From%20Here.
45. Sloan, *Daylighting the Saw Mill River*, 52–53.
46. John Golden, "Carpet Mills Arts District a Winner in State Awards Round," *Westchester and Fairfield County Business Journals*, January 14, 2016.

47. Ernie Garcia, "Yonkers Demolition Making Way for 442 Apartments," *Rockland/Westchester Journal News*, June 17, 2016; Ernie Garcia, "Downtown Yonkers Block Slowly Razed," *Rockland/Westchester Journal News*, September 7, 2016.
48. Funding for this phase came from both economic development and environmental sources, such as Empire State Development, the New York State Department of Environmental Conservation, New York State Environmental Facilities Corporation, and New York State Energy and Research Development Authority. "Saw Mill River Daylighting Phase III Opens Downtown," *Yonkers News*, November 28, 2018.
49. Sloan, *Daylighting the Saw Mill River*, 53–54.
50. Ernie Garcia, "Yonkers' Chicken Island Parking Lot to Become Residential Center, Hotel," *Rockland/Westchester Journal News*, October 29, 2018.
51. "Chicken Island Sold for $16M," *Yonkers Times*, November 11, 2018.
52. Mansuda Arora, "Follow the River, Follow the Money: On Development in Yonkers," *River* (Kingston NY), October 20, 2020.

BIBLIOGRAPHY

Allison, Charles Elmer. *The History of Yonkers*. New York: Wilbur B. Ketcham, 1896.

Bâcle, Caroline, dir. *Lost Rivers*. Montreal: CatBird Productions, 2012.

Barnett, William C. "Introduction." In *City of Lake and Prairie: Chicago's Environmental History*, edited by Kathleen A. Brosnan, Ann Durkin Keating, and William C. Barnett, 3–14. Pittsburgh PA: University of Pittsburgh Press, 2020.

Brown, Henry Collins. *Old Yonkers, 1646–1922: A Page of History*. New York: Valentine's Manual Press, 1922.

Checker, Melissa. *Sustainability Myth: Environmental Gentrification and the Politics of Justice*. New York: NYU Press, 2020.

Cornell, Thomas C. *Some Reminiscences of the Old Philipse Manor House in Yonkers and Its Surroundings*. Yonkers NY: Yonkers Chapter of the Westchester County Historical Society, 1934.

Gibson, Campbell, and Kay Jung. *Historical Census Statistics on Population Totals by Race, 1790 to 1900, and by Hispanic Origin, 1970 to 1990, for Large Cities and Other Urban Places in the United States*. Population Division Working Paper no. 76, U.S. Census Bureau, 2005. https://www.census.gov/content/dam/Census/library/working-papers/2005/demo/POP-twps0076.pdf.

Hall, Edward Hagaman. *Philipse Manor Hall at Yonkers NY: The Site, the Building, and Its Occupants*. New York: American Scenic and Historic Preservation Society, 1912.

Madden, Joseph P. *A Documentary History of Yonkers, New York: Volume Two, Part Two: The Dutch, The English, and an Incorporated American Village, 1609–1860*. Bowie MD: Heritage Books Inc., 1992.

Moffit, Dr. Henry. "Exhibit A: Testimony Taken at Hearing Held in Yonkers, January 22, 1892." In *Thirteenth Annual Report of the State Board of Health of New York*. Albany: State Printer, 1893.

Owen, James. "Old Mills along the Saw Mill River." *Quarterly Bulletin of the Westchester County Historical Society* 6, no. 1 (January 1930): 1–7.

Scharf, J. Thomas. *History of Westchester County, New York, Including Morrisania, Kings Bridge, and West Farms, Which Have Been Annexed to New York City, Volume II*. Philadelphia: L. E. Preston & Co., 1886.

Shonnard, Frederic, and W. W. Spooner. *History of Westchester County, New York: From Its Earliest Settlement to the Year 1900*. New York: New York History Company, 1900.

Sloan, Jennifer. *Daylighting the Saw Mill River: A Curriculum on Green Design for Urban Ecology Enrichment and Resilience*. Yonkers NY: Groundwork Hudson Valley, 2019.

State of New York. *Fortieth Annual Report of the State Board of Health: For the Year Ending December 31, 1919, Volume II: Report of Division of Sanitary Engineering*. Albany: J. B. Lyon Company, 1920.

Stevenson, C. Ian. "Introducing Environmental History into Vernacular Architecture: Considerations from New England's Historic Dams." *Buildings and Landscapes: Journal of the Vernacular Architecture Forum* 24, no. 2 (Fall 2017): 1–21.

Van der Donck, Adriaen. *A Description of New Netherland*. Edited by Charles T. Gehring and William Starna. Translated by Diederik Willem Goedhuys. Lincoln: University of Nebraska Press, 2008.

13 Angling in the Anthropocene

Carp and the Making of Race on the Los Angeles River

Bryan B. Rasmussen

My ideal fly-fishing experience is a few days in California's Sierra Nevada, on any of the hundreds of limpid streams that run through its glacier-carved granite canyons or green-gold meadows rimmed in snow-capped mountains. I have fled the "busy," "discontented" life of the city, and in just a few hours' drive, I am coaxing a "bold and lusty" trout to rise to my artificial fly dead-drifting in the bubble line, or enticing it to scoop up a subaquatic nymph pretending to be trapped in an eddy. Fly-fishing here is not far, culturally speaking, from the pastoral "silver streams" celebrated by Izaak Walton, the seventeenth-century English angler who elevated this "harmless recreation" to an art. The charms of Walton's *The Compleat Angler* (1653), his "watery discourse" on angling, are inseparable from the gentlemanly virtues that he attached to the unspoiled nature of the English countryside.[1] Because of Walton, fly-fishing has "that aristocratic thing that we all secretly love," confesses longtime angler Jim Burns of California's South Coast chapter of Trout Unlimited. To anglers like Jim and me, fly-fishing promises immersion in a certain kind of nature distinctive to lettered men like Walton—a place of "contemplation and quietness," an extension of the gentleman's library, where we "can fish and study too."[2]

Jim echoes Anders Halverson, who writes that it was fly angling's very "pretense to elitism that made it first attractive to Americans." Indeed, the sport that Walton helped to ennoble has actively set itself apart from the more conventional recreation of bait fishing in some key ways. Though

23. Looking north from Los Angeles, the cities of Burbank and Glendale are visible here. The I-5 hugs the river to the left, while the Ventura Freeway (SR 134) crosses the river. The concrete confluence of the Verdugo Wash is visible beneath SR 134. Recycling, HVAC, and autobody industries are visible to the right of the river. Beyond SR 134 is an electric power plant. Photo by Bryan Rasmussen, 2021.

Walton himself fished with all manner of baits, fly anglers fish with artificial "flies" made to look and behave like real insects, rather than with lowly worms or hot dogs. Though Walton catalogued numerous delicious fish recipes, fly anglers today mostly practice catch and release, signaling that they fish because they want to, not because they have to. And while bait anglers can be found dunking worms in any reservoir or off any bridge or culvert, fly anglers traditionally seek out more pellucid waters of pristine landscapes.

Walton's art is the one Tim looks like he's practicing. Tim, a TV editor from Santa Monica, is standing knee-deep in the river in full fly-fishing getup: waders, angler's vest, lacquered wooden trout net, and fly box of artificial bugs that bespeaks his familiarity with fish and their natural history. Tim's bright-hued fly line slices softly through the air, a signature display of angling skill and finesse "not attainable by common capacities." Standing in full sun, framed handsomely by riparian flora, Tim looks like a living embodiment of the recreation Walton celebrated as "somewhat like poetry."[3]

Except Tim isn't on one of Walton's Arcadian streams: he's on the Los Angeles River. If it's possible to have the opposite of an Izaak Walton fly-fishing experience, angling on the LA is it. At its best, the LA is, as poet Angela Peñaredondo describes, a river of "tatted boxcars" and helicopters "whizz[ing] by like dragonflies," "a river broke down to skin and bones."[4] At its worst, it's closer to Luis Rodriguez's "urban-spawned / Stream of muck . . . pouring over nightmares of wakefulness."[5] To angle here, Tim has descended the steep bank of a concrete flood-control channel draped in crackling power lines and circumscribed by a landscape zoned for heavy commerce and industry. He's listening to the roaring, crisscrossing streams of freeway and railroad that cinch the river in asphalt and steel. He's gently guiding his fly between half-submerged shopping carts, invasive *Arundo donax*, and floating garbage, not in expectation of a "bold and lusty" trout, the sport angler's favored fish, but for common carp, maligned species of human-engineered environments.[6]

The striking incongruity of Tim, practicing a recreation first imagined on faraway rivers, and this manufactured landscape is all the more vivid for Oscar, standing just a few yards upstream of Tim. Oscar's spot couldn't be more different. Where Tim has found a passable imitation of Waltonian nature on this post-natural river, Oscar has posted up in the shade under the cool girders of the Glendale Freeway overpass, its concrete geometry filled with treated wastewater and the echo of cars. He waits for a passing carp, hands in pockets, a pair of fishing rods jammed into gaps in the river's artificial conglomerate shore. He points to some torn tortillas strewn on the bank when I ask him what he's fishing with. Oscar, born in Jalisco, México, was raised here in Los Angeles, in Koreatown, by an uncle who taught him to fish on the LA. Oscar looks like he's had it hard: there's a milky-white globe behind a nasty scar where his right eye should be. He comes to the river on weekends to land some carp, which he takes home in a white five-gallon bucket. He says they're best eaten fried. Oscar makes a little fun of Tim hunting up and down the river, trying to get a rise out of a fish by casting repeatedly into the riffles. Oscar thinks Tim lacks patience and doesn't notice that Tim's technique is borrowed from trout fishing.

Tim and Oscar aren't just practicing two kinds of fishing. Though within shouting distance, they are in reality fishing on two different rivers. Tim is fishing on the white river. The white river isn't exactly a place; it's an idea

24. A bait fisherman on the LA River. Photo by Bryan Rasmussen, 2021.

that fly anglers like him (and me, too, if I'm honest) bring to the kinds of places fly anglers have traditionally sought to escape. Walton's pastoral pastime became in the nineteenth-century United States a signature leisure activity of wealthy urban elites with the privilege to fish for fun, not food. Fly angling afforded "an excuse to get away from the city," with its poverty, filth, disease, and, of course, *race*.[7] To this day, fly angling retains, both in the popular imagination and in its demographics, its historical connection to educated white people fleeing to white spaces.[8]

Oscar, meanwhile, is fishing on the minoritized river. Once a free-flowing seasonal stream meandering mostly peacefully through the willow forests of the San Fernando and Los Angeles Basins, the Los Angeles River was repurposed to serve riverside industries. The river's transformation into an economic instrument and, consequently, an environmental hazard, required that industrialists and urban planners quarantine non-whites in racially unequal spaces along the river, in economically uneven roles as cheap labor to white capital. Channelizing the river in concrete starting in 1938 only formalized more than a century of racial and economic inequity by etching it permanently into the landscape.

Fly-fishing for carp on the Los Angeles River entangles anglers with a fish whose (un)natural history parallels the river's remaking and the making of race on the river. The same racialized processes of urbanization and industrialization that minoritized the Los Angeles River and its human inhabitants also minoritized carp. Walton fished for carp, too. Though they were "naturalised" and not native to England, he called them "Queen of the rivers," second only to salmon, the "King of freshwater fish."[9] But carp have fared differently in the United States. First introduced to feed urban immigrant labor, common carp, foremost among urban wildlife, have become proxies for environmental ruin and racism. Epithets like "junk fish" and "trash fish" speak to an enduring disdain for carp, the places they live, the people who share those places with them, and those who rely on them for food.

This braided history of river, people, and fish has new urgency on the Los Angeles River in light of recent revitalization initiatives. Revitalization would break concrete, bring back native species like steelhead trout, and create new recreational opportunities. But revitalization's problems threaten to follow the river's concrete path of racial injustice by inviting wealth and privilege onto a river that whites once eagerly abandoned.

Fly angling has always been more than a recreation. For Walton, angling was equal parts "recreation" and "contemplation." He called his "watery discourse" a "recreation of a recreation." In this essay, I take Walton's reflexive invitation as an opportunity to rethink, rather than repeat, fly angling's history of privilege. I'll use my own watery discourse to contemplate what it means to fly-fish for carp on the LA while white, and to ask what new, more equitable arrangements between the river, people, and fish might point toward a more just future.[10]

Making a Minoritized River

Remaking an entire river in concrete marks an obvious benchmark of the Anthropocene, that epoch (origins up for debate) during which humans have "been laying down in stone"—better yet, concrete—"a durable archive of human impacts to Earth's geophysical and biophysical systems."[11] As Blake Gumprecht writes, "Perhaps nothing symbolizes the role of human beings in changing the face of the earth more than the exploitation and transformation of the Los Angeles River."[12] Designed for a single purpose, the Los Angeles River's concrete chute and army of tributaries ferry practically every drop of rain that falls on three mountain ranges—along with the chemical runoff of every over-watered lawn, driveway, and car wash in every river-adjacent community—along a "water freeway" capable of traveling its fifty-one miles from Encino to Long Beach faster than one can drive, before flushing it into the Pacific.[13]

But as critics have observed, the Anthropocene's pretense of collective responsibility for a damaged planet, as "a global problem" caused universally by "human beings," hides that the "meta-processes" of the Anthropocene—colonialism, capitalism, industrialization, urbanization—are, in fact, the result of a racially uneven "geography of wealth and power."[14] These meta-processes remade the Los Angeles River not just once, in concrete, but three times. They accumulated over time to produce a minoritized river, a landscape of racial inequality that confined non-whites to economically unequal roles as cheap labor to white industry. Their legacy today is carved into the landscape and onto the communities of minoritized people.[15]

A history of the river's minoritization might begin on August 2, 1769, when Spanish missionary Juan Crespí of the Portolá expedition from San Diego to Monterey looked out across the riverine landscape of what is now called Elysian Valley, near the Indian village of Maungna, and saw what he described as "a most beautiful garden." Crespí and his men came looking for a place to settle a pueblo. Their first priority was agriculture. They eyed this "good sized, full flowing river" with its "very good water, pure and fresh," its trees and fertile soil, and saw everything they needed for "a very large and plenteous mission."[16]

To explorers predisposed to European pastoral landscape ideals, this may have looked like a found Eden, but in reality Eden had to be made. To the Native Tongva-Gabrieleno people who lived with the river for more than ten thousand years, the river already possessed its own meaning and function. The river's floodplain forest ecology provided everything a civilization needed to thrive. Gabrieleno ethnobotanist Mark Acuña recalls "the great willow forests" that stretched the length of the river, from its headwaters in the foothills above Encino to its exit into the Pacific at Long Beach. He describes oak forests so dense a squirrel could hop the length of the river without ever touching the ground.[17] Willows gave medicines and building and hunting materials. Oaks gave acorns, a dietary staple of California Indians. Though they did not practice agriculture in the European sense, Indians used burning to cultivate a more productive landscape. Through traditional knowledge, they knew both the river's gifts and its moods.[18] Their largest village, Yangna (meaning "salted earth"), in present-day downtown Los Angeles, near where a Catholic church now stands, sat well above the floodplain and away from a river channel prone to catastrophic flooding.

Spanish agriculture changed the river from a floodplain ecosystem into an agricultural resource, and it changed Indians' relationship to the river from beneficiaries of its biodiversity to indentured laborers to a colonial ideal. Indians under the missions were the principal source of labor for the pueblo. Indians dug the river's *zanjas*—the irrigation ditches that supplied water to riverside farms growing beans, maize, wheat, corn, barley, and eventually grapes for wine. Over the course of a hundred years, river diversions gradually robbed the floodplain forest to realize a Spanish garden created by Indian and, later, Mexican laborers, whose usefulness for agriculture left them permanent residents in riverside communities vulnerable to floods. This was the first remaking of the river.

By the nineteenth century, the Spanish garden had transformed Los Angeles from a small agricultural village that could support a few thousand people into a city poised to support hundreds of thousands. The U.S. acquisition of California in 1850 following the Mexican-American War drew white land developers seeking to profit from the Spanish paradise. Lured by real estate boosters' promises of health and economic opportunity, the city's population exploded from fewer than six thousand in 1870 to more than fifty thousand by 1890.[19] Most of them were white. Most of them came by rail after the arrival of the Southern Pacific Railroad in 1876. If Spanish colonials had transformed the river into an Avalon, corporate rail made that colonial paradise accessible and profitable to white newcomers.

The Southern Pacific connected Los Angeles to San Francisco, but it also linked the river to the larger project to secure the West for the expanding American empire. The two projects were themselves coupled by Chinese labor. In what Ryan Dearinger calls "the paradox of immigrant labor," Chinese fleeing war and famine at home became in the 1860s the chief labor pool on a transcontinental railroad that established California as a white settler state, against "counter-sovereignty" claims by Indians and

Mexican Californios. Like rail, the Chinese were instruments of colonialism and industry, but they experienced virulent resistance to permanent settlement. They were regarded as "a cancer in the biological, social, religious, and political systems," synonymous with "filth" and "contagions."[20] Leland Stanford, railway tycoon, university founder, Republican governor of California, and xenophobe, fretted in a public expression of white supremacist panic that "unless we do something," California would soon be overwhelmed by Asia's "numberless millions." He advocated for the Chinese Exclusion Act (1882) to bar Chinese laborers from citizenship and limit further immigration. But in his private business, he employed the Chinese to bring the Southern Pacific to Los Angeles. More profitable than prejudice was an expendable labor force.[21]

Rail followed the river downtown. So did the Chinese, who settled at the rail's terminus, in the small riverside Chinatown community near present-day Union Station. Chinese-built rail shifted the function of the river from agriculture to industry. Rail "facilitated industrial growth throughout the city" but especially near Chinatown, where manufacturing and materials industries like brick and clay, foundries, packinghouses, oil field suppliers, gas and water works, glass plants, furniture manufacturers, warehouses, and lumber yards proliferated.[22] Entire portions of the river were given over to rail yards themselves, like Southern Pacific's three-mile-long riverfront Taylor Yard north of downtown. (The original Chinatown was later displaced some blocks north to its present location to make way for new tracks and terminal for Southern Pacific.) This was the river's second great remaking.

Rail tightened the connection between race and industry on the river by providing ready access to cheap labor in Asian and Mexican riverside communities. Mexicans had "eked out lives" on the river long before there was a railroad, in places like Sonoratown, near the original pueblo area of downtown closest to the river, adjacent to old Chinatown.[23] The river "occupied a vital role in [Mexican] village life and folkways."[24] In the white imagination, Mexicans' proximity to the river made them "primitives close to nature," indistinguishable from the river itself.[25] Perhaps because of this association, when rail came, Mexicans' fate followed the river's in becoming an economic instrument for the growing city of Los Angeles. Poor Mexicans, including refugees fleeing the 1911 Mexican Revolution, fed riverside industries. There they could find jobs but little opportunity for "occupational mobility"; they could find cheap housing but on river bottomlands prone to flooding and exposed to industrial waste.[26]

Asians and Mexicans, alienated by their "historical and continuing role as low-wage workers," found themselves on an increasingly minoritized river.[27] The river gave the relationship between non-white labor and white capital a spatial as well as economic reality. Eastside communities—barrios lining the river—"existed in a world apart," cut off from the rest of the city by rail and industrial zoning.[28] Segregation deepened whites' racialized view of the river: riverside communities full of "peaceful primitives" became "vectors" of poverty, filth, crime, and race.[29] Elites fled to outlying areas, where they kept neighborhood racial covenants that, along with zoning laws and city transportation policies, kept poor non-whites "quarantined" on the river, permanent sources of cheap labor.[30] The colonialist-capitalist nexus of rail and industry had remade the river into a wasteland. Elite flight created "an urban social system based on a racially segregated geography."[31] It remained only to guarantee that system in concrete.

By the 1930s, the river "had become foul and septic" with refuse from feedlots, stockyards, and slaughterhouses; oil and grease from rail yards; iron, boron, and chromium from manufacturing plants; and untreated sewage festered in the bed of a river whose off-season flows were rarely strong enough to move it out to sea.[32] Meanwhile, white suburban sprawl along upstream tributaries had reduced the landscape's ability to absorb floodwaters during storms, increasing flood risk for minorities living closest to the river.[33] Every rainstorm became a potential disaster.

This fouled and hazardous river had been entirely the creation of white industry and development, and it was largely to preserve those interests that the Army Corps of Engineers remade the river a third time. Floods like those in 1914, 1918, and 1934 saw continental-scale outflows smash industries that had built up too close to this regional-scale river's edge, taking out key infrastructure like railroads and bridges. After a flood in 1938 claimed eighty-seven lives, the city of Los Angeles contracted with the Army Corps to begin a thirty-year project to channelize the river. The Corps' mandate, to move water through the city as efficiently as possible, was "organized and financed" by railroad industrialists and real estate developers who stood to profit from floodplain development. Channelization disguised profit motive behind "hazard zoning."[34]

25. Looking south along I-5 and the Glendale Narrows. Photo by Bryan Rasmussen, 2021.

The details of this hydro-engineering project are remarkable. Encasing an entire watershed in concrete had never been attempted before or since. The result was, in architectural historian Joseph Giovanni's words, a "sun-bleached landscape of concrete," "imperial in scale," recalling the monumentality of ancient Egypt and Rome.[35] The project remains to this day "the largest public works project undertaken west of the Mississippi."[36] But dwelling on an engineering solution to an unnatural hazard as a case study distracts us from the meta-processes of the Anthropocene.[37] Like the Anthropocene concept itself, channelization "obscures" the racialized "geography of wealth and power" that produced this landscape. It obscures that channelization didn't only *produce* conditions of inequality and vulnerability for minoritized communities along the river, it *required* them. The Army Corps may have channelized the river, but that would not have been possible, or even necessary, without the racialization of labor and landscape that had transformed the river into a sacrifice zone for industry and urban development, its denizens into an expendable workforce. The remaking of the river in concrete was a collaboration of capitalism and colonialism begun more than a century before. An opportunity to formalize urban racial social control at scale and permanently inscribe economic inequality on the landscape, channelization was more symptom than cause of the racial Anthropocene.[38]

White Folks Fly-Fishing

The goal of the Army Corps was total. But despite one of the most coordinated assaults on nature the world has ever seen, folks like Tim and Oscar can fish the LA because, somehow, a sliver of wilderness has managed to sprout from a crack in the tectonic plates of the city. In its pre-channel life, the river was mostly an arroyo—a dry riverbed of soft alluvium—for most of the year. The river ran chiefly during rains, and only during the heaviest storms did it make it to the ocean. The Army Corps could easily slice into the soft landscape to install its concrete gutter. Not so in the Glendale Narrows. The granite riverbed and persistent water in this eight-mile "soft-bottom" (that is, unconcreted) stretch from Burbank to downtown defied excavation and refused concrete. Stymied, the Corps lined the river's wide, sloping banks with concrete and stabilized its floor with boulders. The mountain sediment that gathers here has produced substrate for an urban wilderness.

To people like Oscar and the mainly Asian and Latino riverside communities quarantined on the minoritized river, the Narrows mitigates some of the legacy of environmental racism: its ersatz park space helps relieve park poverty; its opportunities for subsistence fishing help alleviate food insecurity.[39] To white fly anglers, the Narrows's braids of picturesque rivulets burbling through willows and bulrushes do the river's best impression of a trout stream. The river becomes "recognizable" here.[40] Less dystopian sci-fi and more like Wordsworth's "sylvan Wye."[41] Standing hips deep in water, navigating a bed of boulders and muck, I can mentally erase power lines and concrete banks and imagine myself on a "river from another, greener coast."[42] If I listen, I can almost hear Walton whisper between the *whoosh* of cars and the sound of moving water: "God never did make a more calm, quiet, innocent recreation than angling."[43]

But while fly-fishing the Narrows offers a refuge—however manmade—of calm and quiet, it is far from innocent, not unlike the history of angling itself. Fly angling's historically aristocratic pretense belongs to the anti-urban, elitist, racist, xenophobic, and exclusionary stream of American conservation history. Its popularity in the United States grew in proportion to nineteenth-century racial anxieties about the influx into cities of Asian and eastern European immigrants and Black migrants, whose labor was instrumental to white urban capital but whose physical vigor posed

26. A great egret fishing for crayfish among the river's myriad shopping carts that indicate the river's role in homing the unhoused. Photo by Bryan Rasmussen, 2021.

27. A characteristic scene on the "soft-bottom" (unconcreted) portion of the river known as the Glendale Narrows, not unlike where Tim was fishing. Photo by Bryan Rasmussen, 2021.

a threat to the "great race"—an idiom for a cultural nationalism defined by white racial supremacy. Sport fishing and other "manly arts" promised racial rejuvenation in cleaner, healthier spaces. Ideas about conservation and nature recreation grew out of white culture's desire to protect such spaces while managing anxieties about racial decline. Elites justified the making of nature preserves and the criminalization of subsistence hunting and fishing through the logic of protecting wildlife from overhunting and overfishing—bywords for exclusion given that subsistence was the principal relationship to nature among Indians, poor non-whites, and immigrants.[44]

While immigrants and minorities were excluded from nature, and in some cases from citizenship, and increasingly quarantined in the cities they were poorly paid to pollute, their labor quite literally made it easier for white capitalists to flee those cities for the exclusive pleasures of the countryside. Rail traffic from anglers seeking escape became so profitable that many rail companies "went to great lengths to advertise the quality of the fishing along their lines." Immigrants may have built the railroads that afforded access to ideal hunting and fishing, but only elites could afford the fare.[45]

Trout—the sportsman's favored fish—represented the true game fish because it reportedly provided the best fishing, but also because it was the nonhuman exemplar of the sport angler's elitist, anti-urban, xenophobic values. Trout, wrote nineteenth-century angling enthusiast Charles Hallock, "inhabit the fairest regions of nature's beautiful domain. They drink only from the purest fountains and subsist upon the choicest food their pellucid streams supply. On nothing does the flavor and general appearance of a trout or a salmon depend so much as the character of the water in which he lives." Hallock, inheritor of Walton's anti-urban aesthetic, baldly contrasted trout's elite white values with the "brutish," "crowded," "groveling" nature of cities, whose "foul or sluggish waters" "demoralize" fish and angler alike.[46]

Making a Minoritized Fish

As trout is to the sportsman, carp is to urban minorities. Carp is the anti-trout. Byproducts of the same "meta-processes" that minoritized the river, carp is an indicator species of the racial Anthropocene. Trafficked across the planet for so long that they are now native to nowhere, carp belong to the colonialist-capitalist global trade in people and other organisms. Their distribution spans every continent but Antarctica, every state in the United States, and "just about every freshwater habitat" in every state. Their ubiquity in U.S. freshwaters owes to their ability to thrive in human-engineered environments "incapable of supporting gamefish," like the LA, but also to the migration into U.S. urban labor markets of ethnic minorities, for whom carp was "a cheap and abundant source of food." As a result, carp, foremost among urban wildlife, have become proxies for places and people minoritized by urbanization and industrialization.[47]

Stories of carp on the LA reinforce its minoritized status. According to "legend," says environmentalist William Bowling for one local fish study, carp were introduced on the river by Chinese rail workers.[48] The legend isn't only wrong; it obscures carp's function as a tool of settler colonialism and white industrial capital (more in a moment). The same legend references "Asian carp," but this, too, is a mistake with a racial bias. Asian carp collectively refers to four species of carp native to eastern Asia, none of which is the common carp (*Cyprinus carpio*). Records show that Asian carps weren't introduced to the United States until the 1970s. The most widespread of the Asian carps and the one reported to live on the Los Angeles River, grass carp (*Ctenopharyngodon idella*) wasn't introduced to California until 1974 or 1975.[49] Given their small numbers (I've never seen one) on the LA relative to the abundant common carp, it's very unlikely that grass carp are the carp people have in mind when they say "Asian carp." The fact that "Asian" has colloquially come to describe all carp, not just problematic ones like *C. idella*, likely owes to historical anti-Asian racism: carp and Asian immigrants have alike been considered "invasive" in the United States.[50] Persistent myths like these consolidate the association of immigrants with carp: both have been blamed for environmental ruin, and both are instruments of cultural denial analogous to the Anthropocene concept itself.

In fact, common carp arrived in California in 1877, imported by the U.S. government, intended as a food fish for immigrant labor, and almost certainly transported on railroads financed by government and industry and built by the Chinese. Stanford, enthusiastic supporter of Chinese labor on the Southern Pacific, also strongly supported importing carp to feed his workers. Carp tended to find purchase at the ends of rail lines, like in Los Angeles where Chinese rail workers settled. Like rail and Chinese labor, carp belonged to a sustained campaign to supplant native human and nonhuman populations. The U.S. Fish Commission, biological flank of U.S. imperialism, promoted carp's racial superiority in vigor and economic value over native fish, which they considered "worthless and unpalatable." Biological colonialism and railway colonialism intersected in U.S. Fish Commission employees themselves, many of whom owned railway stock. Just as the U.S. and California governments colluded with American railway industrialists to use "alien" Chinese labor to displace Indians to make room for whites, they programmatically stocked carp on California's freshwater streams and lakes to replace weaker, economically worthless native fish with alien fish that whites favored.[51]

However, despite a "tremendous explosion" in carp popularity at its first introduction, enthusiasm soon fell sharply owing in part to whites' distaste for carp as a food fish and to its association with immigrants and non-whites.[52] Carp remained popular in cities with large communities of immigrants with long-established traditions of eating it in their home countries.[53] Carp's role as street food reinforced its image "as 'fish of the masses' or 'poor man's fish.'"[54] In the white imagination, carp and people living in degraded urban environments alike were "aliens," "criminals," "'villains' beyond redemption."[55] Wildlife managers' assessments of carp followed these racialized attitudes, driving the cultural prejudice that separated game fish from food fish, and sport fishing from subsistence fishing.

When the bottom fell out of the carp market, fish commissioners quickly pivoted to stocking trout on pristine rivers and selling fishing licenses to anglers with time and money to recreate there. Wildlife agencies that once extolled carp now excoriated them as "biological pollutants."[56] Just as sport anglers and hunters blamed minorities and the poor for overhunting and overfishing on their preserves, wildlife managers blamed carp for stealing space and resources from their preferred game animals. Carp, they now said, outcompete game fish for space and food. They dirty waters where game fish were valued. They eat trout fry and even ducklings. Securing trout streams for quality angling now meant eliminating undesirable fish like carp from waters that agencies had once indiscriminately stocked with them.

In the 1960s, wildlife conservation shifted again, this time to restoring native habitats and species. Anti-carp bias persists among conservation ecologists, but this owes more to social prejudice than scientific fact. For

28. Frogtown and Taylor Yard. Looking northeast across the I-5 (*bottom*). Photo by Bryan Rasmussen, 2021.

just as poor, minoritized communities became proxies for urban degradation, carp have been unjustly blamed for ruining native ecosystems. Far more detrimental to native species than carp are "human alterations of the aquatic environment," whether damming rivers and draining wetland habitats to fill reservoirs or channelizing urban rivers to build on bottomlands.[57] Nevertheless, carp's ability to thrive in places that would kill native species has been used against them, as if any fish that would thrive in degraded habitats was complicit in ecocide. For their crimes they've been condemned as "dumpster dolphin," "ghetto grouper," "mud marlin," "sewer salmon," "Tijuana trout," and so on. Such names reveal less about carp than our relationship to the places where they live and to the people who share those places with them and depend on them for food. In the white imagination, carp became proxies for race and environmental ruin, but really they're a mirror reflecting our values back at us.

Carping for Trout

The history of carp parallels the making of race on the Los Angeles River. The meta-processes of the Anthropocene that minoritized the river simultaneously created a habitat in which carp, monsters of the Anthropocene, could thrive. Fly-fishing for carp—or "carping," as it's called—on the LA imports a recreational ideal saturated in privilege and exclusion to the kind of place white anglers have historically been eager to abandon. It is a nexus of colonialism, capitalism, and conservation that Tim symbolizes perfectly. Tim is knee-deep in one of the most scenic spots on the LA, along the Glendale Narrows in Elysian Valley, in a neighborhood colloquially known as Frogtown, the selfsame site where Crespí first described his pastoral paradise.[58] A bit upstream of Tim is Oscar, on the concrete bank. Directly across from Oscar someone has scrawled "stolen land" on the freeway stanchion beneath the overpass, a helpful reminder to urban tourists like Tim that colonialism here is ongoing.

Tim, a TV editor, drove in from Santa Monica, lured by Instagram images of gleaming golden carp caught on the fly. He admits little experience with carp angling, so he's using strategies learned on more ideal rivers, angling for more ideal fish. He's practicing what I call—because I have done it—carping for trout. Carping for trout is the tell of the white urban fly angler. In the first place, it turns something that should be easy into something "stupidly hard," reports Jim Burns, curator of the longtime *LA River Fly Fishing* blog (subtitled "Fishing for Carp, Waiting for Steelhead"). Trying to catch a carp on an artificial fly guarantees you're there for sport, not subsistence. Moreover, it turns something that should be cheap into something very expensive. Fly-fishing affords the angler the opportunity to lay out hundreds or thousands of dollars on specialized gear: waders and boots, multi-pocketed angler's vest, polarized sunglasses and sun protective clothing, and fly rod. Tim's fly box is jammed with flies he bought for three or four dollars apiece from the Orvis outfitter in Pasadena. Gear like this shows off the angler's eagerness to get into the river, to be one with nature. But given that fly anglers likely lead urban, not rural, lives—like Walton himself—this eagerness to get *into* nature also signals a detachment *from* nature. It marks us as tourists, visitors who do not remain—a relationship to nature distinctive to white cultural norms. Oscar, cool in the shade of the overpass, doesn't feel the need to get *into* the river like Tim does. Maybe that's because he's *already* in it. Like the minoritized communities relegated to the river, maybe Oscar never left.[59]

Carping for trout gives whites the opportunity to imagine a place for ourselves on a minoritized river. It claims space for a white enjoyment that has historically come at the expense of subsistence and economic opportunities for non-whites. Its signature totems convey class and status, passports to anywhere, like whiteness itself. It expresses a desire, however implicit, to remake space in our own image, to bring our ideas of nature anywhere we go. It was this same desire that drove missionaries to remake the river in the image of European pastoralism, white capitalists to remake the river into an industrial service pipe, and the city of Los Angeles to collude with the federal government to concretize the river and its injustices. Now, for perhaps the same reasons that Tim has chosen to recreate in the Narrows, city planners, private developers, and environmentalists have descended on this charismatic portion of the river with promises to remake it a fourth time, but troubling signs point to new injustices that follow the historical ones laid down in concrete.

Frogtown, in particular, is ground zero for revitalization. A historically underserved, largely Latino community on mixed residential and industrial zoned land squeezed between the Interstate 5 and the river, Frogtown sits directly opposite Southern Pacific's Taylor Yard, which in the 1930s was an industrial hub that spanned three miles along the river. The city of Los Angeles purchased its decommissioned G2 Taylor Yard parcel in 2017 and made it a park, though technically it's still a brownfield site. The heavy-metal vestiges of the river's prior industrial uses will need to be cleaned up before the temporary soccer fields can be turned into something more permanent.[60]

Despite this residue of harm, most of the buildings that front the river in Frogtown—relics of some of the only industries where, not that long ago, Asians and Latinos could find work—have been converted into condos, design studios, live-work spaces, and coffeeshops slinging draft lattes. Between 2000 and 2010, whites in Frogtown swelled by more than 100 percent. Since 2010, when revitalization initiatives really got going, whites grew a further 76 percent, minorities remained about the same, and land values more than doubled.[61] Where once riverside communities like Frogtown experienced some of the worst consequences of the river's abuse, they're now experiencing the worst consequences of revitalization in the form of gentrification and displacement.[62] Revitalization will certainly increase recreational opportunities along the river, but it threatens new river enclosures that could "save" the river by securing it for white use. Eighty-five percent of respondents to the first multilingual public opinion poll on river revitalization said they would "likely visit the river if it had more opportunities for recreational and cultural activities," but 68 percent reported concern that "river revitalization may lead to gentrification or displacement of current residents." That jumps to 79 percent among respondents from households earning less than $40,000 per year.[63] Meanwhile, "Studies project the LA River's revitalization will spur billions of dollars in riverside development," writes Amir Khafagy, but many of these development projects are moving ahead *before any significant changes to the river itself*.[64] Developers are speculating on the *idea* of a revitalized river. Like carping for trout on the LA, revitalization is an exercise in who gets to imagine themselves on the river.

Brown Folks Fly-Fishing

What is river revitalization that doesn't remake the river, again, in the image of white desire? How can traditionally white recreations like fly angling point to more just and equitable arrangements between the river, people, and fish? I can't answer these questions alone any more than I could catch a carp alone. The day my research student Colin and I land our first carps on the fly, we do so with a lot of help from Lino Jubilado, long-time angler on the LA. Lino takes Colin and me down early one Sunday to Long Beach, to fish a willowy lagoon where concrete river meets ocean, a borderland at once industrial and primeval. Carp easily in the double digits are breaching all around us with great fat splashes, like aquatic dinosaurs in the morning fog. It doesn't take long for Colin to hook one. He's fishing on my slim fly rod designed for one-pound trout, not ten-pound carp. With Lino's coaching, he's able to play the fish, keep the line taut so it can't get off the hook, but not so taut he snaps the pricey rod in half. Colin wrestles it for almost seven minutes before he gets it close enough for Lino to scoop it up in his huge, aluminum-and-rubber net. Next to the dainty wooden one dangling from my vest, Lino's net appears designed to trawl for ocean debris. Colin needs it: at more than twenty inches, his carp is bigger than any fish I ever caught. While Lino unhooks the fish, Colin shakes his weary rod arm and smiles big, exhausted but elated. Lino

29. Lino (*left*) and Colin angling in Long Beach, one of the soft-bottom portions of the river. Photo by Bryan Rasmussen, 2021.

sets him up for a photo op: Colin balances the rod behind his neck while leaning forward to present the fish with two hands to make it look a little bigger—not that the fish needs the illusion. Lino snaps a few and records Colin giving a little speech to commemorate his first LA River carp on the fly before sending the creature on its way. Lino makes us feel like clients of a pro-river guide on some epic fly-fishing excursion. That's his aim: give regular folks that experience, but for free, and on a wastewater river.

Lino grew up in Alhambra, in East LA. Some of his earliest memories are of fishing for fun and food with his dad and granddad. But Lino attributes his passion for fishing to his uncle, a sport fisherman. Lino always imagined he'd go into policing, like his father, but went into social work instead. After getting his master's at University of Southern California, he went to work in Santa Monica on a housing-first homeless mitigation initiative. Get people into housing so you can treat them. "It's a great model," he recalls. When I ask him about changes he's seen on the river, he remembers the '80s when he'd have to cut a hole in the fencing to fish.

He remembers worrying his car would get broken into. He's upbeat about gentrification: "You've got really nice houses and it's really artsy fartsy. . . . I love that whole vibe."[65]

But maybe Lino's sanguine attitude toward gentrification is also a relic of his past life as a sport angler. He became a competitive bass fisherman after his uncle. He fished competitively for a decade, but he bristled at the territoriality of the sport fishing community. "They think they own the river. . . . They're like, if you [bring people onto the river for free] more people are going to show up." But, he says, "That's exactly what I want!" He knew people who wanted to learn fly-fishing but couldn't "because there's a price tag." "People were like, 'Oh, you should hit up Lino because he'll actually take you down there and treat you like a client and get you on a fish.'" His inclusive ethic has run him afoul of sport anglers because doing it for free has "taken out the whole business end of it." He thinks, "We should be able to share the water." When he brings people to the river, he says, "I feel like I'm actually fishing, and I get more stoked when someone else catches, honest to God."[66]

Lino's got no illusions about the river: "Honestly . . . those are the cards I'm dealt." Sure he loves the Sierra Nevada, with its "wild native stuff," but he thinks you don't have to "get away" to have a nature experience. Lino's shift from the exclusive pleasures and profits of sport fishing to something more inclusive comes from a desire to conserve *this* river, a place he loves. "You get people down here and they get hooked," he says. "They want to participate in the river cleanups. They want to go to the meetings. . . . That's how it starts. You know, you just bring awareness."[67]

Lino's inclusive river ethic extends to carp, too. "I have a lot of history with these fish. . . . It's just a part of me. And they're gorgeous." He lovingly refers to them as "golden nuggets" because "who doesn't like gold?" But his affection for carp speaks to a deeper desire for belonging. We talked about revitalization initiatives like reintroducing the river's native steelhead trout—a fish that hasn't been seen on the river since the start of channelization. To reintroduce trout would almost certainly mean eradicating carp. Lino balks at the idea: "I mean, we've got multiple generations now, carp that have grown up in here thinking, 'This is where I belong.' It is totally their home." It's not hard to imagine he's also talking about himself and his place on the river, and in Los Angeles. Lino is second-generation Filipino.

30. Colin shows off his carp. Photo by Lino Jubilado, 2021.

His grandmother, who recently passed, was the last in the family to speak Ilicano and Tagalog. "None of my family have any accent whatsoever, so it's just, it's gone." His father still makes the traditional foods, but, "You know, we're it. It's been that way for a while. I mean, we're totally Americanized now. America is family."[68]

Another urban fly angler, Fernando Vázquez, also sees himself reflected in carp. Fernando studies industrial design at an area art school. He thinks he might want to design for a fly-fishing gear company one day. Fernando is from Michoacán. That's where his father taught him to fish. He never ate them: "I always had this empathy for animals." He says he doesn't know why "people talk so much crap about carp." People ask him, "Why are you fishing for shit fish?" They "just have a big stigma." But when I probe, he says, "There's a lot of facts about it, but I think there's also, like, a racial" thing. He's "highly against calling a fish 'trash fish' and 'mud marlin.'" He has particular feelings about the term "Asian carp." "If people are already . . . racist enough to call someone by their race, people will follow that and then degrade the name of that fish and degrade the fish. They're just going to hate that fish automatically." His own experiences on the river have mirrored those of carp. Anglers question his abilities or assume "you're not into it because you're Hispanic." Does that happen a lot? "Yeah. Enough for me not to care about it anymore."[69]

Maybe fly-fishing's exclusive, whites-only vibe is why Fernando loves his solitude on the river. On the river, "I don't really run into people, and that's why I like it." He likes the quiet of the Narrows, especially, where "the river kind of comes back to its original form." But the river's surprises include opportunities for human connection, too. "You meet all kinds of people." Professionals, researchers, scientists. "As an angler, I don't care where you're coming from as long as we get along in the fishing world." He's also met "people who are really troubled." He got to talking one day with a "guy from México and his daughter" about fishing. The guy took a real interest, asked to pose with his daughter and Fernando's fly rod. Fernando took pictures. "They were all happy about it." Turns out the guy had been running from the police for two years because he missed a court date. Fernando told him, "'If you're not going to turn yourself in, just do the right thing and be functional to society. And look, you're already in love with fishing and you've never even done it. Get into it. It's going to change your life, you know?' I'm pretty sure he probably picked that up at some basic level. He was really excited." Fernando loves this story. Angling is about repairing relationships: with fish, with people, with ourselves. "People actually need to fish to be okay with themselves," he says. "I feel like everybody needs some sort of outlet for their personal issues," including, for example, personal loss: after his mother died in 2021 from COVID complications back in México, Fernando's carping Instagram took a real uptick.[70]

The Brown River

Revitalization's "watery discourse" right now is over concrete: more concrete, less concrete. Proponents of each focus on changing the river itself. Concretists would hire "starchitects" like Frank Gehry to hide more of the river beneath expensive platform parks. Removalists would daylight the river, making it suitable for native trout.[71] But prioritizing concrete in the revitalization conversation prioritizes the effects rather than the causes of injustice on the river. Whether we add or subtract concrete, we need to ask first, will remaking the river a fourth time make it just? Whose communities will be served by expensive infrastructure projects? What will happen to subsistence anglers like Oscar when steelhead are reintroduced? Will they be able to continue fishing for an endangered trout? Before we change the river again to something we want to see, let's ask, what can we learn from what is actually going on there?

Anglers on the LA help us see beyond concrete, beyond favored fish, beyond the *effects* of the minoritized river to deeper causes. They give us a way to think about revitalization, not first as a change in the river, but in our attitude toward it, toward those who recreate there, and toward those who live there. They give us an alternative to the white river. Let's call it the brown river. If the white river is about imagining ideal rivers with ideal fish, the brown river is about angling where you are. The brown river is less an ideal than a practice for being on the river we're already on. The brown river is practicing human and nonhuman connections, inclusive access, belonging, and repair. The brown river shows us that a racialized river doesn't have to be unjust. It can be just. In some ways it already is.

NOTES

1. Izaak Walton, *The Compleat Angler* (1653; Project Gutenberg, 1996), chapters 1 and 4, https://www.gutenberg.org/cache/epub/683/pg683.html.
2. Jim Burns, in discussion with the author, July 2021; Walton, *The Compleat Angler*, chapters 1, 5.
3. Walton, *The Compleat Angler*, chapters 1, 17.
4. Peñaredondo, *Maroon*, 32–33.
5. Rodriguez, *The Concrete River*, 29.
6. Walton, *The Compleat Angler*, chapter 4.
7. Halverson, *An Entirely Synthetic Fish*, 61.
8. Outdoor Foundation, *2020 Special Report on Fishing*, Alexandria VA, 2020, https://asafishing.org/wp-content/uploads/2020/08/2020-Special-Report-on-Fishing_FINAL_WEB.pdf, 22.
9. Walton, *The Compleat Angler*, chapters 9, 7. "Freshwater fish": Walton is referring to the fact that Atlantic salmon (*Salmo salar*) spawn in freshwater rivers.
10. Walton, *The Compleat Angler*, chapter 1, prologue.
11. Nixon, "The Anthropocene," 2. See also Davis and Todd, "On the Importance of a Date," who argue for dating the Anthropocene to the beginning of the colonial era.
12. Gumprecht, *The Los Angeles River*, 3.
13. Gumprecht, *The Los Angeles River*, 224.
14. Pulido, "Racism and the Anthropocene," 117; see also Davis and Todd, "On the Importance of a Date," 763: "Indeed, many people in the humanities have pointed out the failure of the Anthropocene, as a concept, to adequately account for power relations."
15. Pulido, Sidwai, and Vos, "An Archaeology," 433.
16. Crespí, *A Description of Distant Roads*, xx.
17. "Tongva: Walking Earth Keeper Mark Acuña," KCET, January 24, 2011, https://www.kcet.org/shows/departures/gabrieleno-tongva-mission-indians.
18. Deverell, *Whitewashed Adobe*, 97.
19. Quintana, "Making Do, Making Home," 54.
20. Californios were Mexicans who lived in California prior to the U.S. acquisition of California in 1850. Dearinger, *The Filth of Progress*, 152–56; Karuka, *Empire's Tracks*, 100.
21. Leland Stanford, "Governor's Inaugural Address" (California, January 10, 1862); Karuka, *Empire's Tracks*, 82–83.
22. Quintana, "Making Do, Making Home," 53; Gumprecht, *The Los Angeles River*, 111.
23. Deverell, *Whitewashed Adobe*, 110.
24. Deverell, *Whitewashed Adobe*, 97.
25. Deverell, *Whitewashed Adobe*, 92–93.

26. Pulido, Sidawi, and Vos, "An Archaeology," 430–31; Quintana, "Making Do, Making Home," 53.
27. Pulido, Sidawi, and Vos, "An Archaeology," 435.
28. Deverell, *Whitewashed Adobe*, 92.
29. Deverell, *Whitewashed Adobe*, 108.
30. Quintana, "Making Do, Making Home," 51.
31. Quintana, "Making Do, Making Home," 54.
32. Gumprecht, *The Los Angeles River*, 124.
33. Davis, *Ecology of Fear*, 70.
34. Davis, *Ecology of Fear*, 69, 70.
35. Joseph Giovanni, "Just Subtract Water: The Los Angeles River and a Robert Moses with the Soul of a Jane Jacobs," Los Angeles Review of Books, December 18, 2015, https://lareviewofbooks.org/article/just-subtract-water-the-los-angeles-river-and-a-robert-moses-with-the-soul-of-a-jane-jacobs/.
36. Gumprecht, *The Los Angeles River*, 226.
37. Pulido, "Racism and the Anthropocene," 117.
38. Deverell, *Whitewashed Adobe*, 121.
39. See Post and Carter, "Unnatural Nature," especially 1, 2, 10, 12.
40. Giovanni, "Just Subtract Water."
41. William Wordsworth, "Lines Composed a Few Miles above Tintern Abbey," *Lyrical Ballads* (London: J. & A. Arch, 1798; Project Gutenberg, 2003), 1, 58, https://www.gutenberg.org/files/9622/9622-h/9622-h.htm.
42. Giovanni, "Just Subtract Water."
43. Walton, *The Compleat Angler*, chapter 5.
44. See Halverson, *An Entirely Synthetic Fish*, 67, 69, 74. See also Taylor, *The Rise of the American Conservation Movement*, chapter 3, especially 73–82, and chapter 6, especially 185–88; Pulido, *Environmentalism and Economic Justice*, chapter 4, especially 161–63; Beinart and Coates, *Environment and History*, chapter 2, especially 27–31.
45. Halverson, *An Entirely Synthetic Fish*, 67.
46. Hallock, *The Fishing Tourist*, 25–26.
47. Panek, "Biology and Ecology of Carp," 3, 6, 12. See also Dill and Cordone, *History and Status of Introduced Fishes*, 182.
48. Friends of the Los Angeles River, *State of the River* 3, 87.
49. Dill and Cordone, *History and Status of Introduced Fishes*, 223. All four species of Asian carps were introduced by the U.S. government for aquaculture and biological control.
50. "Fish and Other Aquatic Vertebrates," USDA National Invasive Species Information Center, accessed December 14, 2021, https://www.invasivespeciesinfo.gov/aquatic/fish-and-other-vertebrates. The USDA National Invasive Species Information

Center does not list common carp among its invasive species, observing that "determining the invasiveness of a species depends on a number of local factors, including type of habitat." As Sabrina Drill, Natural Resources Advisor Emeritus for the University of California Cooperative Extension in Los Angeles, observes, invasive "only has a contextual meaning. . . . It's invasive when it's causing environmental harm or economic harm or harm to human health." Sabrina Drill, in discussion with the author, May 2021.

51. Dill and Cordone, *History and Status of Introduced Fishes*, 49, 182, 312, 323.
52. Fritz, "Commercial Fishing for Carp," 26.
53. Fritz, "Commercial Fishing for Carp," 24.
54. Fritz, "Commercial Fishing for Carp," 25.
55. Dill and Cordone, *History and Status of Introduced Fishes*, 23.
56. Dill and Cordone, *History and Status of Introduced Fishes*, 301.
57. Fritz, "Commercial Fishing for Carp," 27, 29; see also Dill and Cordone, *History and Status of Introduced Fishes*, 319–20, 327–28.
58. In a past life, Frogtown was habitat for the endemic California red-legged frog.
59. Burns, discussion. See, for example, Norgaard, *Salmon and Acorns*, especially chapter 4.
60. "U.S. EPA Announces $1.1 Million in Brownfields Grants to Promote Economic Redevelopment in Los Angeles," EPA, June 5, 2019, https://www.epa.gov/news releases/us-epa-announces-11-million-brownfields-grants-promote-economic-redevelopment-los.
61. Thanks to California Lutheran University political science student Nakiessa Abbassi for this data.
62. See, for examples, Jon Christensen, "Green Gentrification on the Los Angeles River: A Battle for the Future of the City," KCET, February 20, 2018, https://www.kcet.org/shows/earth-focus/green-gentrification-on-the-los-angeles-river-a-battle-for-the-future-of-the-city; Jonathan Hahn, "Efforts to Restore the Los Angeles River Collide with a Gentrifying City," *Sierra Club Magazine*, June 29, 2018, https://www.sierraclub.org/sierra/2018-4-july-august/feature/efforts-restore-los-angeles-river-California-collide-gentrifying-city.
63. StudyLA, *Public Opinion on the Los Angeles River*, Loyola Marymount University, summer 2021, https://lmu.app.box.com/s/21qn43megmheadmrzgiabslt8e06tm5a.
64. Amir Khafagy, "Gentrifying the Los Angeles River," *Jacobin*, May 16, 2019, https://www.jacobinmag.com/2018/05/los-angeles-river-gentrification-revitalization-displacement.
65. Lino Jubilado, in discussion with the author, July 2021.
66. Jubilado, discussion.
67. Jubilado, discussion.
68. Jubilado, discussion.

69. Fernando Vázquez, in discussion with the author, August 2021.

70. Vázquez, discussion.

71. See Louis Sahagún, "Frank Gehry's Bold Plan to Upgrade the LA River," *Los Angeles Times*, January 11, 2021, https://www.latimes.com/environment/story/2021-01-11/frank-gehry-plan-los-angeles-river; Oliver Wainwright, "Is Frank Gehry Really the Right Person to Revitalise the Los Angeles River?," *Guardian*, October 23, 2015, https://www.theguardian.com/cities/2015/oct/23/frank-gehry-revitalise-los-angeles-la-river; Marissa Christiansen, "We Have One Chance to Save a Crucial Portion of the LA River. Let's Not Blow It," *Los Angeles Times*, July 30, 2019, https://www.latimes.com/opinion/story/2019-07-29/los-angeles-river-taylor-yard-park-project.

BIBLIOGRAPHY

Beinart, William, and Peter Coates. *Environment and History: The Taming of Nature in the USA and South Africa*. New York: Routledge, 1995.

Crespí, Juan. *A Description of Distant Roads: Original Journals of the First Expedition into California, 1769–1770*. Edited and translated by Alan K. Brown. San Diego: San Diego State University Press, 2001.

Davis, Heather, and Zoe Todd. "On the Importance of a Date, or Decolonizing the Anthropocene." *ACME: An International Journal for Critical Geographies* 16, no. 4 (December 2017): 761–80.

Davis, Mike. *Ecology of Fear: Los Angeles and the Imagination of Disaster*. New York: Vintage, 1999.

Dearinger, Ryan. *The Filth of Progress: Immigrants, Americans, and the Building of Canals and Railroads in the West*. Oakland: University of California Press, 2015.

Deverell, William. *Whitewashed Adobe: The Rise of Los Angeles and the Remaking of Its Mexican Past*. Berkeley: University of California Press, 2004.

Dill, William A., and Almo J. Cordone. *History and Status of Introduced Fishes in California, 1871–1996*. San Diego: University of California Press, 1997.

Friends of the Los Angeles River. *State of the River 3: The Long Beach Fish Study*. Los Angeles: FOLAR, 2016.

Fritz, Arnold W. "Commercial Fishing for Carp." In *Carp in North America*, edited by E. L. Cooper, 17–30. Bethesda MD: American Fisheries Society, 1987.

Gumprecht, Blake. *The Los Angeles River: Its Life, Death, and Possible Rebirth*. Baltimore: Johns Hopkins University Press, 2001.

Hallock, Charles. *The Fishing Tourist*. New York: Harper & Brothers, 1873.

Halverson, Anders. *An Entirely Synthetic Fish: How Rainbow Trout Beguiled America and Overran the World*. New Haven: Yale University Press, 2011.

Karuka, Manu. *Empire's Tracks: Indigenous Nations, Chinese Workers, and the Transcontinental Railroad*. Oakland: University of California Press, 2019.

Nixon, Rob. "The Anthropocene: The Promise and Pitfalls of an Epochal Idea." In *Future Remains: A Cabinet of Curiosities for the Anthropocene*, edited by Gregg Mitman, Marco Armiero, and Robert S. Emmett, 1–18. Chicago: University of Chicago Press, 2018.

Norgaard, Kari Marie. *Salmon and Acorns Feed Our People: Colonialism, Nature, and Social Action*. New Brunswick NJ: Rutgers University Press, 2019.

Panek, Frank M. "Biology and Ecology of Carp." In *Carp in North America*, edited by E. L. Cooper, 1–16. Bethesda MD: American Fisheries Society, 1987.

Peñaredondo, Angela. *Maroon*. San Bernardino CA: Jamii Publishing, 2015.

Post, Jason, and Perry Carter. "Unnatural Nature: Anglers' Reimaginings of the Los Angeles River." *Geographical Review* 112, no. 2 (January 2021): 207–27.

Pulido, Laura. *Environmentalism and Economic Justice: Two Chicano Cases from the Southwest*. Tucson: University of Arizona Press, 1996.

———. "Racism and the Anthropocene." In *Future Remains: A Cabinet of Curiosities for the Anthropocene*, edited by Gregg Mitman, Marco Armiero, and Robert S. Emmett, 116–28. Chicago: University of Chicago Press, 2018.

Pulido, Laura, Steve Sidawi, and Robert O. Vos. "An Archaeology of Environmental Racism in Los Angeles." *Urban Geography* 17, no. 5 (1996): 419–39.

Quintana, Isabela Seong Leong. "Making Do, Making Home: Borders and the Worlds of Chinatown and Sonoratown in Early Twentieth-Century Los Angeles." *Journal of Urban History* 41, no. 1 (July 2014): 47–74.

Rodriguez, Luis. *The Concrete River*. Evanston IL: Northwestern University Press, 1991.

Taylor, Dorceta E. *The Rise of the American Conservation Movement: Power, Privilege, and Environmental Protection*. Durham: Duke University Press, 2016.

Contributors

SHAWN BAILEY has a PhD in environmental history from the University of Montana, where he researched and fished the headwaters of the Columbia River. He teaches history at the Illinois Mathematics and Science Academy and lives in Chicago with his wife and son.

ROBERT BARTLETT, PhD, is a retired public sociologist and a storyteller, avid fly fisherman, and lover of rivers. A fifth-generation African American, Bob grew up in West Virginia on a tributary of the Potomac River.

GREG GORDON was born at the junction of Cherry Creek and the South Fork of the Platte River and spent much of his adult life living along the Clark Fork and Dearborn Rivers. He now lives a short walk from the confluence of Hangman Creek and the Spokane River and is a professor of environmental studies at Gonzaga University.

MARGO HILL, JD, MURP, is a Spokane tribal member and grew up on the Spokane Indian Reservation. She serves as the associate director of Small, Urban, Rural, and Tribal Center on Mobility (SURTCOM). Hill earned her Juris Doctor from Gonzaga School of Law and her master's degree in urban and regional planning from Eastern Washington University. She served as the Spokane tribal attorney for ten years and as a Coeur d'Alene tribal court judge and currently teaches at Eastern Washington University.

HEIDI LASHER began floating western rivers on inner tubes, gradually moving on to rubber rafts, plastic kayaks, and wooden canoes. She holds master's degrees in public administration and nonfiction writing. She spends her time writing, working as a communications consultant, parenting teenagers, and scheming her next river trip. She lives in Spokane, Washington.

CHAR MILLER is the W. M. Keck Professor of Environmental Analysis and History at Pomona College, located in the Wanawna/Santa Ana River watershed on Tongva lands in Southern California. He is the author of *Natural Consequences: Intimate Essays for a Planet in Peril* (2022), *West Side Rising: How San Antonio's 1921 Flood Devastated a City and Sparked the Rise of a Latino Environmental Justice Movement* (2021), *San Antonio: A Tricentennial History* (2018), and editor of *On the Border: An Environmental History of San Antonio* (2001, 2005).

BRIAN M. MURPHY, PhD, is passionate about working on the ground with local communities, providing technical assistance and innovative solutions that support leaders working to ensure their communities and rivers remain healthy and resilient into the future. He has worked in the water resources engineering and river science field for more than twenty years as a licensed professional engineer. Born and raised in Spokane, Washington, he lives and works in Denver, Colorado, on Arapaho and Cheyenne lands.

BRYAN B. RASMUSSEN is a professor of English and affiliated faculty in environmental studies at California Lutheran University. He writes about conflicts between conservation and environmental justice at the urban-wildland interface. He is a contributing science writer for the Robert T. Moore Lab's Mexican Bird Resurvey Project, a certified California Naturalist, and a volunteer educator for the Friends of the Los Angeles River.

MARIAN L. RICE grew up exploring the watersheds in the Cascades of Oregon. As a result of her love of the outdoors, both her MPA and PhD research at the University of Utah focused on the interplay of energy extraction and environmental justice in Utah. Combining her experience as a practitioner, she is interested in research on environmental justice, contaminants of concern, and resiliency to climate change.

KIRSTEN RUDESTAM has a PhD in environmental sociology from the University of California, Santa Cruz. Her current research focuses on the dynamics of contested and emergent groundwater use practices within California's Central Coast.

JENNIFER STEVENS is the director of the Arts and History Department at the city of Boise and is affiliated faculty in urban studies and history at Boise State University. She studies and writes about the urban environment in the West and lives with her husband in Boise, Idaho.

C. IAN STEVENSON is the director of advocacy for Greater Portland Landmarks, a nonprofit preservation organization in Portland, Maine. Ian holds a PhD in American and New England studies from Boston University. A scholar of the built environment and environmental history, Ian has written about dams, summer-vacation architecture, and railroads. Originally from Yonkers, New York, he now lives with his wife and son on Peaks Island in Maine.

CHRISTIAN C. YOUNG, PhD (history of science and technology, University of Minnesota), is the director of Conservation and Environmental Science and an adjunct professor of history at the University of Wisconsin–Milwaukee. He taught biology, natural history, and environmental studies at Alverno College in Milwaukee for more than twenty years and is an active volunteer with the Urban Ecology Center.

Index

Page numbers in italics indicate illustrations.

Milton Keynes UK
Ingram Content Group UK Ltd.
UKHW022219200724
445785UK00001B/7

9 781496 230614